UNCERTAINTY AND QUALITY IN SCIENCE FOR POLICY

THEORY AND DECISION LIBRARY

General Editors: W. Leinfellner and G. Eberlein

Series A: Philosophy and Methodology of the Social Sciences
Editors: W. Leinfellner (Technical University of Vienna)
G. Eberlein (Technical University of Munich)

Series B: Mathematical and Statistical Methods
Editor: H. Skala (University of Paderborn)

Series C: Game Theory, Mathematical Programming and
Operations Research
Editor: S. H. Tijs (University of Nijmegen)

Series D: System Theory, Knowledge Engineering and Problem Solving
Editor: W. Janko (University of Economics, Vienna)

SERIES A: **PHILOSOPHY AND METHODOLOGY OF THE SOCIAL SCIENCES**
Volume 15

Editors: W. Leinfellner (Technical University of Vienna)
G. Eberlein (Technical University of Munich)

Editorial Board

M. Bunge (Montreal), J. S. Coleman (Chicago), M. Dogan (Paris), J. Elster (Oslo), L. Kern (Munich), I. Levi (New York), R. Mattessich (Vancouver), A. Rapoport (Toronto), A. Sen (Oxford), R. Tuomela (Helsinki), A. Tversky (Stanford).

Scope

This series deals with the foundations, the general methodology and the criteria, goals and purpose of the social sciences. The emphasis in the new Series A will be on well-argued, thoroughly analytical rather than advanced mathematical treatments. In this context, particular attention will be paid to game and decision theory and general philosophical topics from mathematics, psychology and economics, such as game theory, voting and welfare theory, with applications to political science, sociology, law and ethics.

The titles published in this series are listed at the end of this volume.

UNCERTAINTY AND QUALITY
IN
SCIENCE FOR POLICY

by

SILVIO O. FUNTOWICZ
Joint Research Center, Commission of the European Communities,
Ispra, Italy

and

JEROME R. RAVETZ
The Research Methods
Consultancy Ltd., London, U.K.

Library of Congress Cataloging-in-Publication Data

Funtowicz, Silvio O.
    Uncertainty and quality in science for policy / Silvio O.
Funtowicz and Jerome R. Ravetz.
        p.   cm.
    Includes bibliographical references.
    ISBN 0-7923-0799-2 (alk. paper)
    1. Science and state. 2. Uncertainty (Information theory)
3. Mathematics--Philosophy.   4. Science--Philosophy.   I. Ravetz,
Jerome R.   II. Title.
Q125.F87 1990
338.926--dc20   90-38778

ISBN 0-7923-0799-2

Published by Kluwer Academic Publishers,
P.O. Box 17, 3300 AA Dordrecht, The Netherlands.

Kluwer Academic Publishers incorporates
the publishing programmes of
D. Reidel, Martinus Nijhoff, Dr W. Junk and MTP Press.

Sold and distributed in the U.S.A. and Canada
by Kluwer Academic Publishers,
101 Philip Drive, Norwell, MA 02061, U.S.A.

In all other countries, sold and distributed
by Kluwer Academic Publishers Group,
P.O. Box 322, 3300 AH Dordrecht, The Netherlands.

Printed on acid-free paper

Printed in the Netherlands

To Mirta, Jonatan and Luciana
and also to
Joe, Anna and Tom

ACKNOWLEDGEMENTS

This book is about the NUSAP National Scheme. Its philosophy and basic structure were developed between 1982 and 1985 at the Department of Philosophy of the University of Leeds. The research was supported by the SERC–ESRC Joint Committee of the United Kingdom.

Our analysis of quality was stimulated by work done for the Commonwealth Science Council of London. Chapter 11 is based on research done for the World Resources Institute of Washington D.C. Chapter 12 was done in collaboration with S. M. Macgill of Leeds University; this study was supported by the National Radiological Protection Board of the U.K.

Chapter 10, Section 7, and Chapter 13, Sections 3 and 4, reproduce research done with Chris Hope of Cambridge University. This last study was supported b? the ESRC-Environmental Issues Initiative. In Chapter 13, Sections 1 and 2 are based on research done with Bob Costanza of Maryland University.

We also recall fruitful discussions with Larry Brownstein of the Department of Sociology (University of Leeds).

TABLE OF CONTENTS

PROLOGUE

There is a long tradition in public affairs which assumes that solutions to policy issues should, and can, be determined by "the facts" expressed in quantitative form. But such quantitative information, either as particular inputs to decision-making or as general purpose statistics, is itself becoming increasingly problematic and afflicted by severe uncertainty. Previously it was assumed that Science provided "hard facts" in numerical form, in contrast to the 'soft', interest-driven, value-laden determinants of politics. Now, policy-makers increasingly need to make "hard" decisions, choosing between conflicting options, using scientific information that is irremediably "soft".

Policies can no longer be assumed to be based on scientific information that is endowed with a high degree of certainty. Does this mean that policy-related science can provide only information of low quality? Will such scientific information have any use other than as a rhetorical device for justifying decisions already made on other grounds? We shall show that uncertainty does not necessarily mean low quality in scientific information in policy contexts. Our guiding principle is that high quality does not require the elimination of uncertainty, but rather its effective management. We provide methods whereby uncertainty can be managed, in such a way that users of information can assess its strength relevant to their purposes. These methods also provide means for the expression of uncertainty of the information in the forms best suited to its functions.

The issue of quality-control in science, technology and decision-making is now appreciated as urgent and threatening. The experiences of Chernobyl and Challenger, both resulting from lapses of quality-control, illustrate this problem. We have described the "Ch-Ch Syndrome": the catastrophic collapse of sophisticated mega-technologies resulting from political pressure, incompetence and cover-ups (Ravetz *et al.*, 1986). The destructive impact of our industrial system on the natural environment is another manifestation of the syndrome. Here the phenomena are less dramatic but more pervasive. The pathologies of the industrial system are transferred out, so that it degrades its environment while running "normally". This contradiction affects more than particular high technologies; the very place of science in our civilization is called into question.

We are not among those who deny the reality, objectivity or value of scientific knowledge. But we shall demonstrate that ignorance and error interact with knowledge and power more intimately than was ever conceived hitherto. This is the lesson of the "Ch-Ch Syndrome" for policy making. Approaches to managing our technologies must be based on coping with ignorance at least as much as on the application of knowledge. The exercise of the moral commitments of prudence and integrity is essential to the proper

1

conduct of scientific expertise, whether in quality-control, accident prevention, or the protection of the environment.

Science, seen as knowledge performing special social functions, must
change rapidly in the light of the "Ch-Ch Syndrome". This does not mean that
our civilization will, should, or can abandon science. The situation is analogous to the late-medieval world described by Eco in *The Name of the Rose* (Eco,
1984). Then theology and its associated erudition had lost their inspiration.
The decay was reflected in a local disaster: the monastery fire that started in
the library. In our late-modern world, science and mega-technology have
produced Challenger and Chernobyl. In our transition period as in Eco's the
future structures of belief and of power are scarcely discernible. But a change
in the symbolic and social roles of science is inevitable.

Our approach is intended as a contribution to this larger issue, focussed on
quantitative science; this is the area which has been paradigmatic for the
image of science as the means to absolute certainty. Accordingly judgements
about "quality" in any sense have been considered redundant in relation to
numerical expressions. Quantitative facts have seemed to express assured
truths, and any residual uncertainties were considered to be removed by the
mathematical techniques of statistics. This present work of ours is only a part
of a much larger effort of reconstruction; but we believe that it is central, for
practice and philosophy alike.

As the "Ch-Ch Syndrome" shows, quality-assurance cannot be taken for
granted even in the biggest and most prestigious technologies. Bad workman-
ship, debasing materials and information alike, can persist, unknown to the
public, for decades. The problems of quality in relation to materials are now
familiar, and can be attacked with a concentration of resources and commitment. But quality of information is a new problem scarcely recognized as yet.
The faith in "hard facts" still persists, among publics, decision-makers and
experts alike.

Hitherto the means for establishing quality-control of information have
been fragmentary and specialized, suffering from incomprehension and neglect. Our approach enables a systematic evaluative criticism of the quality
of scientific information. It accomplishes this by fostering the craft skills of the
management of uncertainty. In this way it contributes to the resolution on the
contradiction between the "hard" policy decisions and the "soft" scientific
facts on which they depend.

INTRODUCTION: SOME ILLUSTRATIVE EXAMPLES

This book is about a new approach to the problems of uncertainty and quality of scientific information in the policy context. Its core is the NUSAP notational scheme; the name is an acronym of the initials of an ordered set of categories, starting with Numeral. Their meaning will soon be made clear. We focus on information in quantitative form since this is the traditional means of communication in science. The NUSAP approach is based on a philosophy of mathematics and of knowledge which corresponds to good scientific practice, although not as yet to the consensus in the philosophy of science.

To help to explain what NUSAP is and can do, we will start with a couple of examples that show how uncertainty and quality are real issues in the management of quantitative information. For our first example, we will show how uncertainty and variability are managed in the expression of fundamental physical constants. This might seem paradoxical, since a physical constant should be constant by definition. But the successive "recommended values" of any constant are not all the same; the graph of values of the "fine-structure constant" (Figure 1) is typical. Of particular significance is the way that the "jumps" between successive recommended values are generally greater than the error bars of each such value. It is as if the calculated random error includes only the lesser part of the real uncertainty associated with the recommended value. Until the value of the constant "settles down", the systematic error seems to be at least double that of the random error, every time.

With NUSAP wa can express this double uncertainty quite neatly. In the NUSAP scheme, after Numeral and Unit, we have the uncertainty categories Spread and Assessment. Here we can use these to express the random error and the systematic error, respectively. To emphasize the point, we express both Spread and Assessment in terms of the standard deviation (error bar). Thus the entry in Spread is ± 1 by definition, and the entry for the systematic error in Assessment can be compared directly. The partial NUSAP expression for the recommended value for 1968 is then:

$$\alpha^{-1}(1968) = 137.0 + 360 : E\text{-}4 : \pm 1 : \pm 2\frac{1}{2}$$

By means of this notation, we show clearly and concisely that at that time, the constant had not yet settled down.

The NUSAP notation includes one more category, which, unlike the others, has no immediate analogue in standard statistical practice. This we call "Pedigree". It expresses what its name implies: what sort of background history there is to the number, by which we may be assured of its quality. For simplicity we write the Pedigree as a set of codes (here done on the numerical scale 0–4); in the case of this physical constant it is (4, 4, 3, 4). The "3" is a

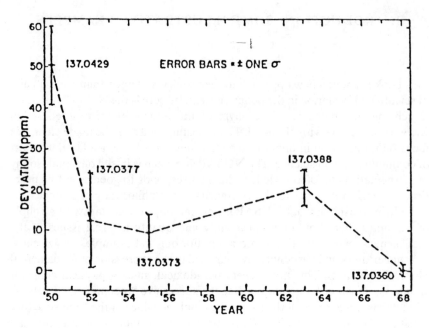

Fig. 1.   Successive recommended values of the fine-structure constand α^{-1} (B. N. Taylor *et al.*, 1969, 7)

reminder that the value is not fully accepted by colleagues, since they will expect it to jump about for a while longer.

Our next example is taken from a recent study in the social sciences. It shows how a set of related estimates of uncertainty can be expressed clearly and effectively by NUSAP. Suppose that we wish to forecast what the future price of a basic commodity might be, especially when at the moment its price is artificially maintained by a cartel of producers. There is no experimental evidence on such a future contingency, and yet we are not completely in the dark. There is a long history of expertise in the field; and there is a well-tried standard model by which experts' guesses can be translated into mathematical form.

The issue was the subject of a research project in the early 1980's, when the OPEC cartel on oil had kept the price high for the previous decade. Clearly oil firms and policy makers were interested in knowing how the price would change in the event of the cartel breaking. The sharp rise of 1974 might not be reversed, because of the fears of depletion of reserves. The researchers surveyed the literature for the data on the prices of oil under the relevant conditions, modified by some informed estimates of their own. They fed all these into a mathematical model, which performed an appropriate number of simulations, and yielded a set of probability distributions. From the mean values ot these obtained a set of estimates of a future competitive price of oil. These varied from $2 to $18 per barrel.

What was to be done with this information? The simplest thing would be to calculate an average, and let that represent the forecast. But that would tell users very little indeed about the distribution; and by itself it would convey no indication of the quality of the result. Although all the estimates were in a sense guesses, they were based on professional experience; they were not the sort of opinions that would be collected by an outside-broadcast TV team on a busy street. The NUSAP scheme provides a compact and easily understood expression for such forecast. Moreover, by its means some subtle distinctions can be made, so that the information can be tailored for a variety of policy problems.

The distribution of forecasts was graphed (Figure 9); in prose, its shape can be described as follows: "In a perfectly competitive non-cartelized market, there is a 10 % chance that the price of oil would be below about $3 per barrel, a 50 % chance of being less than $6 per barrel, a 90 % chance of not exceeding about $11 per barrel".

For a simple NUSAP representation, we take an average, which is 6, at the 50 % mark on the graph. Such a number does not tell a user about the range of values; consulting again the figure, we see that the cumulative graph climbs rather steeply. A convenient way to describe this is to say that 80 % of the distribution lies between 3 and 12. This represents a "factor of 2" on either side of 6; and so we may write in NUSAP, using " : " to separate the entries of the categories

$$6 : US\$: f2 : 80\%$$

The average is the general purpose representative of the distribution; but there are others. For example, an important question, is what it approaches at its upper end. To make this expression meaningful we must take the value at some percentile of the total distribution, say the 95th, for the highest simulation result might well be an outlier of no significance. This 95th percentile number can be read off the graph, as 14. (Why we round up or down to the nearest integer will become clear later in the next, when we discuss appropriate levels of precision). To "flesh out" this number, the researchers examined the process whereby it was obtained; the computer simulations, 100 in number, give a 25 % mathematical sampling error. Hence a NUSAP representation for this estimate is:

$$14 : US\$: \pm 25\% : \%95$$

where we invert the percentage symbol to indicate that percentile is denoted.

So far these numbers do not tell us about their inherent quality. For that we use the Pedigree category. In this case, it has the code (3, 2, 2, 2), indicating that its quality is around the medium grade, and expressing the information: "The results were originally produced using a mathematical model that might not command universal agreement among experts, operating on data that represented only the authors' subjective probability distributions".

These two examples have shown how NUSAP can illuminate good practice in the management of quantitative information in natural and social sciences alike. Equally important, NUSAP can serve as a tool for the analysis, criticism and remedying of bad practice in the handling of numbers. This is far more widespread than is commonly realised; indeed, an important part of the problem of improving quality assurance of information, is that so few people even realise that the problem exists.

There will be cases where the quantitative information is so defective that few if any meaningful conclusions can be drawn from it. NUSAP can then be used as an instrument to detect what we can call "GIGO-sciences", those where the uncertainties of the inputs must be suppressed lest the outputs become completely indeterminate. What proportion of our policy decisions depend on such GIGO-sciences, is a question to be answered by those experienced in those areas; but it is not negligible. A contribution that NUSAP can make in the longer term is the provision of discipline and quality assurance in this important and seriously neglected domain.

CHAPTER 1

SCIENCE FOR POLICY: UNCERTAINTY AND QUALITY

We start with a paradox in the situation of science today. Amidst all the great progress in scientific theory and in technological developments, we are confronted by a new class of environmental challenges and threats. Among these are hazardous wastes, greenhouse effect and ozone depletion. These give rise to problems of a different sort from those of traditional science, either in laboratory, classroom, or industry. Science was previously understood as achieving ever greater certainty in our knowledge and control of the natural world; now it is seen as coping with increasing uncertainties in these urgent environmental issues. A new role for scientists will involve the management of these crucial uncertainties; therein lies the task of quality assurance of the scientific information provided for decision making.

The new scientific issues have common features the distinguish them from traditional scientific problems. They are global in scale and long-term in their impact. Quantitative data on their effects, and even data for baselines of "undisturbed" systems, are radically inadequate. The phenomena being novel, complex and variable, are themselves not well understood. Science cannot always provide well-founded theories based on experiments, for explanation and prediction. It can frequently achieve at best only mathematical models and computer simulations, which are essentially untestable. In this way, it is "soft" scientific information which serves as inputs to the "hard" policy decisions on risks and environmental issues. We can no longer maintain the traditional assumption that certainty is guaranteed by a mathematical language for science. The tasks of quality assurance of scientific information require a new methodology based on a new philosophical foundation.

1.1. INFORMATION FOR POLICY-RELATED RESEARCH

The issues of risk and environment impose new tasks on those scientists and experts who provide information and advice on policy problems. This is the area of "policy-related research". Because of the complexity and urgency of these issues the research communities do not always possess the knowledge and skills required for immediate scientific solutions which are effective for policy. Even experienced advisors may find it difficult to convey to policy-makers an accurate reflection of the scope and limits of the scientific results that can be achieved under these constraints. Policy-makers tend to expect straightforward information as inputs to their decision making process; they want their numbers to provide certainty. But the issues concerning policy-related research involve much uncertainty, and also inescapable social and ethical aspects. Simplicity and precision in predictions, or even in assignment

7

of "safe limits", are not feasible in most cases. Yet they are expected by decision-makers and publics alike. The tasks of properly presenting scientific information and assuring its quality, and also of identifying meaningless numerical assertions, have thus become of great importance for public policy in the areas of risks and the environment.

The character of the information in policy-related research is well described in the press release for a report on Risk Assessment in the Federal Government, prepared by a Committee of the Research Council,

"The basic problem with risk assessment is not its administrative setting, but rather the sparseness and uncertainty of the scientific knowledge of the health hazards addressed", concluded the committee. While "evidence of health effects of a few chemicals, such as asbestos, has been clear, in many cases the evidence is meager and indirect," said the committee. Often available evidence consists entirely of data from animal testing with no direct information on human health effects. "To make judgements amid such uncertainty, risk assessors must rely on a series of assumptions". (U.S. National Research Council, 1983.)

As an illustration provided by the report, the significance of benign tumours in animal tests as indicators of carcinogenicity is unknown, and hence one governmental agency could decide to include such tumours in calculating cancer risk, while another agency could exclude them. Similarly, in the absence of effective dose-response data for a particular substance's hazardous affects, any one of several possible models, some much more conservative than others, could be used to calculate health risks. In such cases, the available information is of inadequate quality in relation to its function as an input to a policy process.

A striking example of this inadequacy is provided by A.S. Whittemore in her important paper on facts and values in risk assessments. The substance in question, ethylene dibromide (EDB) had been tested on rats at three acute doses of 10, 20 and 40 ppm. The incidence of tumours was about 80 % at each of the dose levels. For extrapolation back to 0.015 ppm, the chronic dose for a lifetime risk, three different dose-response curves were employed, giving calculated results, in cases per thousand, of 1, 395 and 551. These were the scientific inputs to a policy decision on the use of EDB to control the Mediterranean fruit fly, in which the stakes were immediate losses of hundreds of millions of dollars per year to the industry, against possible health hazards of up to a million workers. The proposed ambient air standards which were negotiated were 20 and 0.015 ppm (which vary by twice as much as the extrapolated test results), with an eventual compromise at 0.13 ppm (Whittemore, 1983, 25).

The effect of such uncertainties on policy-formation has been graphically described by H.S. Brown in connection with the different approaches to the management of hazardous waste sites.

It can be seen that, despite the similarities in defining cleanup levels for hazardous waste sites, the differences in applying the general concepts are vast. The confusion in terminology, although frustrating, is the least of the problem. The most serious differences stem from variation in the basic assumptions about the environmental fate of chemicals, stringency of application of prin-

ciples of toxicology, data base, use of existing standards/guidelines, use of safety factors, intercon-
version between routes of human exposure, acceptability of cancer risk, and extent of reliance
on expert judgement. Because of this diversity, acceptable ambient concentrations derived by one
method are not comparable to those from another. Furthermore, adoption of numbers derived
through one method for use by another is inappropriate. (Brown, 1987, 258.)

Policy analysts have long been aware of this problem, and have searched for
means of expressing strongly uncertain information. Thus,

One of the thorniest problems facing the policy analyst is posed by the situation where, for a
significant segment of his study, there is unsatisfactory information. The deficiency can be with
respect to data – incomplete or faulty – or more seriously with respect to the model or theory
– again either incomplete or insufficiently verified. This situation is probably the norm rather than
a rare occurrence. (Dalkey, 1969.)

The same awareness can be found among reflective researchers in many fields
of policy-related research. Thus for ecological modelling, R. Costanza and H.
Daly say:

The only solution to the partial quantification trap is to recognize and deal with the range of
imprecision inherent in any decision. This means looking at the full range of possible outcomes,
given the level of precision of our models and data, and making decisions in that context.
(Costanza and Daly, 1987, 5–6.)

Similar problems affect the provision of statistical information, either for
specialised tasks or for general-purpose use. Published statistical information
is the outcome of a complex process, involving a variety of agencies with
differing degrees of mutual communication and control. The quality of the
numbers which emerge is strongly influenced by a number of factors, including: the ways in which the observable entities (environmental, social, economic) are defined and categorized; the standard operational procedures for data
collection and analysis; the sorts of relations between the different levels and
branches of the bureaucracies involved in the work; and the provisions for
monitoring at various phases of the process. Arrays of tabulated numbers,
each expressed to four or more digits, do not convey the degree to which such
quantitative information may be theoretically and socially constructed. In
practice the quality of such information is affected by other factors, including
the costs (in money, aggravation and delays) of obtaining access to it, and
possible restrictions on its use. Finally, when such statistical information lacks
quality assurance (being of unknown provenance or of dubious integrity), it
quality, as an input to a decision process or debate, is seriously impaired.

1.2. HOW TO COPE WITH UNCERTAINTY?

The traditional assumption of the certainty of all quantitative scientific information has become recognized as unrealistic and counterproductive. The
different sorts of uncertainty must be capable of representation. The task was
well described by W.D. Ruckelshaus, when he was Administrator at the US
Environmental Protection Agency:

First, we must insist on risk calculations being expressed as distributions of estimates and not as magic numbers that can be manipulated without regard to what they really mean. We must try to display more realistic estimates of risk to show a range of probabilities. To help to do this we need tools for quantifying and ordering sources of uncertainty and for putting them in perspective. (Ruckelshaus; 1984.)

The above reference to 'magic numbers' is not merely rhetorical. Our culture invests a quality of real truth in numbers, analogous to the way in which other cultures believe in the magical powers of names. The classic statement is by Lord Kelvin,

I often say that when you can measure what you are speaking about, and express it in numbers, you know something about it; but when you cannot measure it, when you cannot express it in numbers, your knowledge is of a meagre and unsatisfactory kind. (Mackay, 1977.)

In this tradition, quantitative assertions are not merely considered necessary for a subject to be scientific; they are also generally believed to be sufficient. Thus the problems discussed here are not only related to the inherent un-certainties of the subject matter (as for example in risks and environmental pollutants); they originate in an appropriate conception of the power and meaning of numbers in relation to the natural and social worlds. By their form, numbers convey precision; an 'uncertain quantity' seem as much a contra-diction in terms as an 'incorrect fact'. But this image must be corrected and enriched if we are to grow out of the reliance on magic numbers; only in that way can we hope to provide useful knowledge for policy-decisions, including those for science, technology and the environment.

Some experts in the relevant sciences, and in the theory of decision-making, hold fast to the old faith. There will never be a shortage of quantitative data, however dubious, nor mathematical models, however abstract and truncated. For research to continue and contracts to flow all it needs is the conviction that with next year's generation of computers we will solve all those problems. But when the simple faith is lost, there is a danger of a collapse into despair. Then the core of rationality and objectivity in science and in the political process is abandoned, with dangerous consequences for politics and science alike. Some authors have even argued as if 'pollution is in the nose of the beholder', and reduce all environmental debates to a conflict between 'sensi-ble' and 'sectarian' lifestyles (see, for example, Douglas and Wildavsky, 1982; for a criticism of this approach, see Funtowicz and Ravetz, 1985). Fortunate-ly, some scholars have shown the way to an enriched conception of the policy process, invoking rhetoric in argument and craft skills in analysis, and thereby protecting the genuine roles of science and reason in public affairs (Majone, 1989). This particular analysis could function as an explanation of the need for NUSAP, from the perspective of the policy process.

The outcome of all these developments is that those who use quantitative information in the policy process are discovering that something is wrong. Numerical information is capable of seriously misleading those who use it. Data expressed as a lengthy string of digits presents a spurious appearance of accuracy. When it is recognized as hyper-precise, so that the fourth and fifth digits are patently meaningless, the question comes up: how many of the digits, if indeed any, are meaningful at all? The expression itself does not tell us; the places for thousands and for thousandths of the same unit have the same standing in our arithmetical notations. When such numbers, affected to an unknown degree by hyper-precision or even pseudo-precision, are injected into policy debates, issues that are already complex and difficult are further confused. They have been many attempts to supplement the existing notations by special codes, but they have been partial and specialized. The problems of management of uncertainty, and hence control of quality, are as yet unresolved.

1.3. DILEMMAS FOR SCIENCE

Issues of uncertainty, and, closely related, those of quality of information, are involved whenever policy-related research is utilized in the policy process. As these issues are new, we do not yet possess practical skills for dealing with them effectively. The simplest, and still most common response of both the decision-makers and the public is to demand at least the appearance of certainty. The scientific advisors are under severe pressure to provide a single number, regardless of their private reservations. The decision-makers' attitude may be typified by the statement of a U.S. Food and Drug Administration Commissioner, 'I'm looking for a clean bill of health, not a wishy-washy, iffy answer on cyclamates' (Whittemore, 1983, 28). Of course, when an issue is already polarized, such simplicity does not achieve its desired ends. With such contested policy numbers in play, debates may combine the hyper-sophistication of scholastic disputations with the ferocity of sectarian politics. The scientific inputs then have the paradoxical property of promising objectivity and certainty by their form, but producing only greater contention by their substance. (See, for example, Nelkin, 1979.)

In the disputes on environmental and occupational hazards, which are bound to increase greatly before they abate, popular conceptions of science tend to change drastically from naive trust to embittered cynicism. Having been told in school, in the media, and by all the accredited experts, that science (in legitimate hands) can and will solve all our technical problems, citizens may then have a different sort of experience, frequently involving procrastination, prevarication or even concealment and deception at the hands of the accredited experts, perhaps even those employed to protect them against their hazards. The sense of disillusion is well conveyed in the title of the article, 'The use of Science in Government – Don't Bother Me with the Facts' (Lester, 1989). Now uncertainty is politicized, as the supposed competence and independence of experts is compromised.

Even when the uncertainties are publicly recognized, the problems of those who represent science are not resolved. Indeed, in many respects the scientist may be faced with an impossible task, once uncertainty has been accepted as significant. For example, in risk assessments, the scientific advisor knows that a prediction like a one-in-a-million chance of a serious accident should be qualified by statements of different sorts of uncertainty, so as to caution any user about the limits of reliability of the numerical statements. If these are expressed in mathematical form the statement is quite incomprehensible to the lay user; if in prose, it is tedious and unclear; and if they are omitted altogether, the advisor can be accused of conveying a certainty that is not warranted by the facts.

A more complex dilemma, affecting science as a whole, is encountered whenever scientists are required to give advice on policy-related issues. Such advice is usually supported by the present or expected behaviour of some critical indicator. *Nature* has stressed this dilemma for science in a recent article. 'Half-truths make sense (almost)'. This was a comment on a prediction of the consequences of the greenhouse effect, using the rise in global mean temperature by 2030 A.D. as the indicator. In such a case any definite advice is liable to go wrong: a prediction of danger will appear alarmist (as 'Cassandra') if nothing happens in the short run; while a reassurance can be condemned (as 'Pangloss') if it retrospectively turns out to be incorrect. Thus the credibility of science, based for so long on the supposed certainty of its conclusions, is endangered by *any* sort of scientific advice on such inherently uncertain issues. Faced with this possibility, in such situations the scientific advisor may prudently decline to provide definite advice when requested by policy-makers. But then science will be seen as not performing its public functions of offering advice when needed, and its legitimacy is threatened. Is there no way through the horns of this dilemma, in which the credibility and the legitimacy of science are both at risk? As *Nature* puts it, 'These are among the trials with which policy research centres must contend. Tell the people that there is a muddle, or give them a clear message that they must man the barricades?' The World Resources Institute's solution to the dilemma, the adoption of a computer model, is described by *Nature* as a 'cop-out' (Maddox, 1987).

Some distinguished commentators have questioned whether computer models should be used at all, in the study of the global environmental problems. In a discussion of the work of the International Institute for Applied Systems Analysis ((IIASA), the American mathematician S. Mac Lane describes 'system analysis' as,

the construction of massive imaginary future 'scenarios' with elaborate equations for quantitative 'models' which combine to provide predictions or projections (gloomy or otherwise), but which cannot be verified by checking against objective facts. Instead, IIASA studies often proceed by combining in series a number of such unverified models, feeding the output of one of such model as input into another equally unverified model... Such studies as these are speculations without empirical check and so cannot count as science... The current efforts at IIASA may be 'state of the art'..., but the 'art' in question involves no real element of science (Mac Lane, 1988, 1144).

In their replay, Brooks and McDonald accused Mac Lane of suggesting that 'we should close our doors, that it is a waste of money to apply research results... to issues on the public agenda' (Brooks and MacDonald, 1988, 496). Supporting them, N. Keyfitz reminds us that,

many of the most difficult problems we have to face cannot even precisely formulated in the present state of knowledge, let alone solved by existing techniques of science...
   Such models, although unsatisfying to many scientists, are still the best guide to policy (*sic.*) that we have.
   ... IIASA was established in the belief that science can contribute to the development of tools to examine and hopefully deal with these societal problems. (Keyfitz, 1988, 496.)

In his final replay, Mac Lane continued to doubt that the global problems should be tackled by making models 'that in the first instance are not verifiable', and adds, 'problems are not solved and science is not helped by unfounded speculation about unverifiable models'. His concluding comment is on quality assurance: 'one essential difficulty with IIASA is that it does not appear to have an adequate critical mechanism, by discipline or by report review.' (Mac Lane, 1988, 1624.)

It is clear that these dilemmas of computer modelling in policy-related research cannot be resolved at the technical level alone. Noone claims that the computer models are fully adequate tools; and yet nothing better is in sight. The critics basically judge them by the standards of traditional mathematical-experimental science, and of course in those terms they are nearly vacuous. Their defenders advocate them on the grounds that they are the best possible, without appreciating how very different are the new sciences of clean-up and survival, in respect to their complex uncertainties, their new criteria of quality and their socio-political involvements. The need is for exceptionally dedicated efforts for the management of uncertainty, the assurance of quality, and the fostering of the skills necessary for both these tasks.

1.4. QUALITY ASSURANCE AND POLICY

We now find ourselves in a situation where major decisions, on the most complex and uncertain issues, must frequently be made under conditions of urgency. To the extent that such decisions are based at least partly on research results and are not ruled by simple power-politics, they depend on the relevant scientific information. What is the quality of that information? Does anyone expect industrial systems to run effectively without programmes of quality assurance on inputs, processes and products? It is a commonplace that major industrial disasters are caused by exceptional defects in quality control, rather than by the laws of chance or acts of God. This, after all, is what the 'Ch-Ch Syndrome' is all about: the failure of those responsible for our material well-being to maintain complete control of the technological system. It is well known that bureaucratic social systems can on occasion lose all contact with reality, partly through the neglect (perhaps deliberate) of quality assurance of their information inputs. The system can mask its increasing incoherences until it suddenly falls apart. Yet in the management of our natural environ-

ment, as well as an important areas of our social activities, there is little or
no conception of quality assurance of information.

Another reason for the lack of quality assurance of information derives from
our dominant conception of knowledge. Traditionally 'knowing-that' something is true by some abstract argument has been considered far superior to
'knowing-how' to make something. The categories of 'knowing-that' admit
only simple oppositions like true and false. Judgements of quality may be
subtle and complex, and based on craft skills, all derived from 'knowing-how'
experience. Nearly all influential philosophers have concentrated on arguments that attempted to provide general proofs or disproofs of the possibility
on the attainment of truth. Those who studied the complex interaction of truth
and error, of knowledge and ignorance, requiring an appreciation of 'knowing-
how', have been neglected or misinterpreted. Francis Bacon, for example, had
a very shrewd and penetrating understanding of the causes of error. He had
a general framework for the frailties of the intellect, which he called the 'Idols'
of the Cave (individual), Tribe (culture), Marketplace (common language) and
Theatre (education, especially higher). Furthermore, he was keenly aware of
the ubiquity of error, and of the delusions of confidence. On the theme of
quality, he offered the aphorism: 'What in observation is loose and vague, is
in information deceptive and treacherous' (Bacon, 1621, 98). But such insights
as these were ignored by his optimistic followers, and he was remembered
mainly for the 'positive' message of his laws of induction.

Now quality of information is back on the agenda, and in a context full of
conflicts and confusions. It can no longer be left to the personal wisdom of
the occasional reflective philosophers, writing for those few among posterity
with whom their message may strike a chord. The survival of our civilization,
through its response to the growing environmental threats, will depend critically on the quality of its scientific inputs. Are these to be left to custom,
chance and luck; or is there to be a recognition that policy, no less than
hardware, needs quality assurance of inputs, processes and products? If we
believe that the environmental challenges and threats are the leading contradiction for our civilization, and that effective policy-related research is essential
for our meeting these challenges, then we must agree that the problem of
quality assurance, through the management of uncertainty, is crucial.

To some extent we have been misled by our successes. We have been told
for many years that the big problem with information is its sheer *quantity*. So
much data or 'knowledge' is being produced all the time, that ever more
powerful and elaborate automatic systems are required for its storage and
possible retrieval. In all this technologizing, there have been only occasional
queries as to the quality of the materials being generated and processed. Yet
as we have seen, whenever there is an urgent issue for resolution, it is quality
rather than quantity that presents the problem. Still, how much easier to
propose still bigger and better data-management systems, than to undertake
the arduous task of initiating quality-control. In fairness, this neglect can in
part be ascribed to the absence hitherto of effective tools, conceptual and

methodological. The NUSAP scheme is intended to remedy that defect, and thereby contribute to the resolution of the larger issues.

1.5. UNCERTAINTY AND POLICY

Since our discussion concerns information to be used in a policy context, we must keep in mind that policy-makers have their own agendas; and these can include the manipulation of uncertainty in various ways. Procrastination is as real a policy option as any other, and indeed one that is traditionally favoured in bureaucracies; and 'inadequate information' is the best excuse for delay. More generally, those who operate in a political context may attempt to influence the ways in which their statements and actions are perceived and evaluated. This involves affecting public attitudes, controlling the flows of information and misinformation, and setting the agenda and terms for debate on major issues. Now that uncertainty has been politicizied, as an accepted element of issues of public concern, it too will be manipulated. Parties in a policy debate will invoke uncertainty in their arguments selectively, for their own advantage. The danger with this situation is that there are at present no mechanisms towards a consensus on such politicized uncertainties. Being generally vague in their form and content, they are not amenable to criticism and correction. Where there is no disciplined dialogue, a meeting of minds, or even acquiescences, are less likely; and policy debates will tend to be seen as pure power politics.

Hence the management of uncertainty in policy-related research is an urgent task. The task is not impossible, once the problems are analyzed. First, it involves a translation from the uncertainty (mainly cognitive, but with a value component), to the assessed evaluation of decision options (in which commitments and stakes are the primary focus). Then, provided that there can be some degree of consensus, it can be possible to see how robust is the specification of the problem, in relation to the uncertainties in its inputs. A good example of this has been given by M. Thompson and M. Warburton, in relation to deforestation in the Himalayas. Starting with the paradoxical 'fact' that estimates of the per capita fuelwood consumption vary through a factor of 67, they show that all serious studies agree that their numerical predictions imply that the problem exists and is urgent (Thompson and Warburton, 1985). Attempts to achieve precision in quantitative estimates would be costly, and probably fruitless, and in last resort, irrelevant. Achieving such a sensible solution, even in this extreme case was not easy; but it was done. More common, however, are the cases where the uncertainties balance or even swamp the available knowledge. Even then it may be possible to classify the elements of the decision to be taken with respect to the underlying uncertainties. Thus for the greenhouse effect, there could be scenarios of climate change with their consequences for habitation, agriculture and industry; and these could be keyed to dates for decision, and the structure of the relevant information at those times.

We do not claim that all policy issues can be resolved by such a combination of good science and good will. First, the problem may be so far developed as to be irreversible, and the only practical questions are those of coping, for example, with environmental conditions vastly changed for the worse. Or perhaps the issue, while not yet known to be irreversible, is too complex, and the uncertainties totally swamp the relevant facts. One sign of this situation is a regular succession of discoveries of new active elements strongly affecting the phenomena, along with radically new theories. However, it is possible that even when the available science can provide information that is adequate to its policy function (with uncertainty not eliminated but properly managed), various vested interests may inhibit, delay or distort any public debate. These can include, for example, sectors of industry that would bear the cost of change, as well as politicians with their own agendas, and even institutions of established science, to say nothing of NIMBY groups concerned solely to ensure that it is 'Not in *My* Back Yard'. For each of these actors, uncertainty may be used to justify their own positions. The 'quality' of the information, as seen by each of them will depend on their intended use, and that is conditioned by their commitments, practical and ideological. Any genuine attempt to improve the quality of scientific information as it is used in the policy process must be undertaken with such political realities in mind.

The most important overriding political reality of all, in relation to quality of information, is that now there happens not to be 'someone in control' of quality. With a level of quality assurance of information that would soon reduce any industrial system to ruins, many of our inputs for policy-related research are at the mercy of every sort of manipulation, mystification and power-politics. No technical device or methodological insights can solve this issue unaided. But they can at least serve to expose the secret that that the problem of quality of information exists; that it is constantly with us, and not merely when a Challenger or a Chernobyl explodes; and finally that with a new conception of scientific information in the policy context, it can be managed.

CHAPTER 2

UNCERTAINTY AND ITS MANAGEMENT

Uncertainty has always been a factor in public affairs. The application of science to reduce this uncertainty extends back at least as far as the state astrology practiced by the Babylonians. They carefully kept records of good and bad natural events, and attempted to establish inductive generalizations of their correlation with celestial phenomena. Thereby they hoped to ascertain the judgements of their gods on their activities. With the development of the modern scientific world-view, causality was conceived as impersonal, and uncertainty came to be analyzed in terms of regularly recurring events. Out of this shift came our current ideas of 'probability' in the seventeenth century. Since then the techniques of statistics have developed continuously, enabling an effective management of uncertainty over an increasing range of problems.

As natural science has grown and matured over recent centuries, it has developed tools for the management of different kinds of uncertainty. Each particular set of tools was devised in response to a recognized problem. Quantitative measurements have been made since antiquity in such fields as astronomy; but not until the early nineteenth century was an effective "calculus of errors" created. In a separate tradition, "combinatorial probabilities" were created in the seventeenth century for the analysis of games of chance. These mathematical tools were then available in the later nineteenth century for use in all natural sciences involving random processes. A parallel development was in "statistics" from the seventeenth century onwards, involving aggregated information gathered for its importance to statecraft and commerce. These three approaches can be seen to relate to different aspects of the limits of scientific knowledge: "errors" relate to the limits of exactness of measurements made with real instruments; "randomness" relates to the limits of causality and determinism as observable in the natural world; and "statistics" relates (implicitly in its practice) to the limits of correspondence between descriptive categories and the reality to which they refer. These three approaches have all interacted with and enriched each other, so that now they are not seen as distinct in name or subject-matter.

Another way of looking at the history of management of uncertainty is in terms of the relation between the researcher and the system under study. Combinatorial probabilities describe an abstract world where knowledge of processes is incomplete; but where events occur totally independently of the observer and have no imprecision in themselves. The theory of errors arose from the realization of the complex interaction among the different elements of the measurement process. These include the instruments whose fineness of scale and accuracy of construction and calibration is limited, and the human operators with their individual inaccuracies and distortions of perception. In statistics there is an analogous effect; general concepts (such as "population", "income") must have operational definitions; and data collection and analysis must have safeguards against a variety of possible errors.

The new issues of policy-related research have revealed new problems of uncertainty, for which the classical methods are inadequate. One response to this problem has been to re-emphasize a more "personal" conception of probability, as "degree of belief" (Keynes, 1921) or "betting odds" (Savage, 1954); and on that basis to apply the formalisms derived from Bayes' theorem. This enables the application of mathematical techniques to problems where the data are too sparse or weak to support traditional frequential or combinatorial probabilities and their calculi. Another extension of the realm of applicability of mathematical techniques has been achieved by "fuzzy set" theory, where vagueness is quantified (Zadeh, 1965). These new methods have had their natural use in the quantification of risks, most notably the "major hazard" of large, complex and novel industrial installations (For a comprehensive survey, see Stephanou and Sage, 1987). The integrating scientific field here is PRA (Probabilistic Risk Assessment), in which such techniques, allied with computer simulations, are employed to analyze (and quantify) the structure of those hazards.

2.1. UNCERTAINTY IN PROBABILITY

Probabilistic risk analysis is a significant example of policy-related research. Although its form is that of physical science, it has been oriented and shaped by very definite policy problems, most notably that of the safety (and hence acceptability) of civil nuclear power. Also, it provides a good illustration of the strengths and limits of the probabilistic approach to the management of uncertainty. This includes both the scientific function of risk assessment, and also the political function of reassurance. This latter requires techniques of communication of ideas that are novel, abstract and paradoxical, to an inexpert and perhaps suspicious public.

Such were the tasks undertaken in the Reactor Safety Study, whose results were published as the "Rassmussen Report" (WASH-1400). One of the most severe difficulties experienced by the analysts was in the estimation of some of the probabilities to be used as inputs for the computer simulations. For example, the failure-rates for components had to be considered as constant throughout their lifetimes; and data on possible complex contingencies was simply lacking. In the report were such predictions as the following: "The most likely core accident would occur on the average of one every 17,000 years per plant"; and "The likelihood of being killed in any one year in a reactor accident is one chance in 300,000,000" (U.S. Nuclear Regulatory Commission, 1975). These probability statements did not convey any of the sorts of uncertainties involved in their production (about these uncertainties, see, for example, Rivard et al., 1984), and the report was immediately subjected to very strong criticism (Lewis et al., 1978). Had there been a serious attempt to communicate these qualifying attributes of the quantitative information, the WASH-1400 report might have had a useful function in the furthering of the debate on methods of assessing nuclear reactor safety. But, with the rejection of its overly precise numerical conclusions came a discrediting of its naive probabilistic approach.

Since then some practitioners of Probabilistic Risk Analysis have tried to convey the inevitable inexactness of their method of obtaining quantified probabilities for disasters. Thus,

Uncertainties in estimates of probabilities of events by factors of less than two or three can hardly be expected, and uncertainties by a factor of ten or more may well occur, even in carefully conducted studies. The estimation of the magnitude of the consequences in human terms almost always involves environmental modelling and similar factors of uncertainty are to be expected. (Dunster and Vinck, 1979.)

When calculations compound several numbers with uncertainties of a factor of 2.5 or 10, the resulting uncertainties will be far greater than those of which the public is generally aware.

Even when historic data are available, the categories in which they are cast may render them less than fully relevant to the actual problem at hand, and leave a severe residual uncertainty. A good illustration of this is given by accidents with the marine transport of Liquified Natural Gas (LNG). In a paper on this subject, W.B. Fairley lists the possible sources of uncertainty in: the collection of statistics, with reporting errors; extrapolation; speculatively based estimates; definitional error; and unsubstantiated theory. To show how definitional error may vitiate the statistics, he cites the official claim that there have been no "serious marine LNG accidents in the U.S.A.", when in fact there have been serious accidents that were: marine with other gas; LNG but land-based; and marine LNG abroad (Fairley, 1977). The same example is cited in a report by the Royal Society on risk assessment, to illustrate pseudo-precision in risk calculations. They quote an official American report on the probability of a LNG disaster killing about 100,000 people. This is given as 5×10^{-50} per year; its reliability is given as odds of greater than 10^{40} to 1. They comment that this exemplifies the aphorism attributed to Gauss, "lack of mathematical culture is revealed nowhere so conspicuously as in meaningless precision in numerical computations" (The Royal Society, 1983).

Attempts to resolve such problems of statistical uncertainty by elaborated formal techniques, which might be called "multi-storey" statistics, can become quite baroque. Carrying debate to ever higher levels of method and methodology, they are converted to the status of a weapon in struggles over environmental policy. For example, if the "best estimate" of the probability of a catastrophic event is 10^{-6} yr^{-1}, but with the upper 95 % confidence level it rises to 10^{-4}, which value of the probabilities should be accepted for policy purposes? If objectors demanded near-perfect certainty in the experts' judgements, then over 99 % confidence could be required; the probabilities would then be well up in the "danger" zone. Which level of confidence is appropriate? Given the state of the foundations of probability and statistical theory, it is doubtful that refined calculations on such problems can achieve any genuine scientific goals, as distinct from polemical advantage.

Indeed, there is now an increasing tendency for public debate to focus more on the various uncertainties surrounding the numbers than on the policy-relevant quantities themselves. Even the judgements of the quality of such numbers become crucial elements of debates on their implications. Such debates on the uncertainties will always be inherently more difficult to control and to comprehend than those at the policy level. They unavoidably involve all aspects of the issue, from state-of-the-art expert practice in the relevant scientific fields, to policy and even to methodology. There is a need for methods whereby such debates can be disciplined and guided.

2.2. STATISTICS, COMPUTERS AND UNCERTAINTY

Of course, the inexactness of quantitative data and the unreliability of the inferences based upon them, have long been recognized in the theory and practice of statistics. Over the years, techniques have been developed which impose ever less restrictive assumptions on the data, and which still produce useful tests of hypotheses. The present question is how effective are such techniques when the "sparseness and uncertainty" of the scientific inputs are severe. If a data-base consists of three scattered items, each with an inexactness of a factor of two or more, what meaning can there be in a test of an hypothesis at any level of significance? When a small set of data on acute effects of a toxicant on test animals must be extrapolated for a "safe limit" of chronic doses to humans, how much extra certainty can be achieved by refined statistical tests or computer simulations? Such problems distinguish the new policy-related research from the traditional laboratory, industrial and field situations for which the standard statistical techniques were first developed. This is the area which primarily concerns us; where statistical methods can be effectively applied, our approach can also assist in the evaluation of data, parameters and models, and in the interpretation and representation of results. But there are many problems in policy-related research, including the most urgent among them, where the uncertainties are too great to be effectively managed by techniques deriving from those other contexts.

Our approach should be seen as complementary to the statistical methods as commonly taught and practiced. It can be a useful supplement in cases where the data are appropriate for manipulation by standard techniques; but it is essential in the circumstances of policy-related research. As J.C. Bailar puts it:

All the statistical algebra and all the statistical computations are of value only to the extent that they add to the process of inference. Often they do not aid in making sound inferences; indeed they may work the other way, and in my experience that is because the kinds of random variability we see in the big problems of the day tend to be small relative to other uncertainties. This is true, for example, for data on poverty or unemployment; international trade; agricultural production; an basic measures of human health and survival.

Closer to home, random variability – the stuff p-values and confidence limits, is simply swamped by other kind of uncertainties in assessing the health risks of chemical exposures; or tracking the movement of an environmental contaminant, or predicting the effects of human activities on global temperature or the ozone layer. It was, in fact, this aspect of environmental problems that first attracted me to the field. I have long had an interest in non-random variability, and here I see it in almost pure form. (Bailar, 1988, 19–22.)

The examples given in the above quotation are all in the fields of policy-related research, with which we are primarily concerned. In these, the uncertainty of scientific information is experienced as a problem of the quality of the inputs to decision processes. What Bailar describes as "non-random variability" is the sort of uncertainty for which our approach has been developed.

As computers of ever greater power become accessible, their use in policy-related research increases steadily. We are only now emerging from the period when it was widely believed that because of their digital logic, computers are perfectly exact and error-free. Even when computers are used for strictly mathematical operations, as in the solution of differential equations, which are independent of data-inputs or physical interpretations, this "numerical analysis" requires skills and judgements of several sorts. Thus,

Barring blunders (that is, machine malfunctions, erroneous programming, and other human errors), errors will arise from the fact that continuous variables have been made discrete, that infinite mathematical expressions or processes have been made finite or truncated, and that a computing machine does not do arithmetic with infinite exactitude but with, say, eight figures. Numerical analysis attempts to make an error analysis for each algorithm. (Davis and Hersh, 1981, 1885.)

Error analysis is a highly skilled task. First it involves the management of the inexactness of digital operations, including the design exercise of defining appropriate bounds for error arising from truncation and rounding-off, and setting their limits. Criteria for "goodness" of a (necessarily approximate) answer must also be chosen from among all the alternatives. For the reliability of the work, attention must first be paid to possible blunders and their management. More design comes into the choice among algorithms, related to their reliability and utility in different circumstances. When computers are used for providing inputs to policy-problems, the compounding of the successive differences between scientific theories, mathematical models, computer algorithms, input data, observed phenomena, and the aspect of the real world impacting unfavourably on ourselves, creates a situation which can be seen as including major areas of ignorance along with knowledge.

2.3. TYPES OF UNCERTAINTY

In its unavoidable reliance in computer models and simulations, policy-related research in all fields is concerned with sorts of uncertainty which are only now being recognized. For example, in radiological assessments, scientists talk about three main sources of uncertainty: "data uncertainties", "modelling uncertainties" and "completeness uncertainties". Data uncertainties are those which arise from the quality or appropriateness of the data used as inputs to models. Modelling uncertainties are those arising from an incomplete understanding of the modelled phenomena, or from numerical approximations that are used in mathematical representations of processes. Completeness uncertainties refer to all omissions due to lack knowledge. There are techniques for reducing data and modelling uncertainties, but completeness uncertainties are, in principle, unquantifiable and irreducible (Vesely and Rasmuson, 1984). Clearly, ignorance is involved in the completeness uncertainties; but it may also be present in the other sources of uncertainty mentioned, as, for example, when data are sparse or when model parameters are based on data with little relevance to the actual problem. These sources of uncertainty require craft skills for their management; the completeness uncertainties require judgements concerning the total problem and the degree to which a model can simulate the phenomena and their causes.

A similar analysis is provided in a report prepared for the United States Nuclear Regulatory Commission. Thus,

Because we can only model, not measure, severe-accident frequencies and consequences, the results of PRA's – Probabilistic Risk Assessment – are subject to a variety of uncertainties. Uncertainties in model inputs (e.g. component failure frequencies, initial and boundary conditions, and material properties) will cause model outputs to be uncertain. Uncertainties or approximations in the modelling of individual processes will also contribute to uncertainty in the model outputs. Interrelationships between pieces of equipment or among various processes may not be well understood or properly modeled. Finally, there is always the question of completeness: Have we identified al of the important accident-initiators, equipment failure modes, and physical phenomena. (Rivard *et al*, 1984, 1/4.)

The same report also discusses the increasingly popular activity of classifying uncertainties. Thus,

... from "data uncertainties" to "interpretations uncertainties" (doubtfulness or vagueness in the interpretation of the results). Certainly... the... two general categories of "experimental uncertainty" (variation in results in repeated experiments) and "knowledge uncertainties" (lack of knowledge yielding vagueness, indefiniteness, or imprecision in an analysis, a stated conclusion, or a stated value) represent an important distinction. (Rivard *et al.*, 1984, 1/8.)

Another classification is provided by M.C.G. Hall; his sources of uncertainty are: process; model; statistical; and "forcing", involved in predictions which presuppose values that are unknowable (Hall, 1985, 340). Attempting to manage uncertainties in terms of their separate sources are liable to introduce more problems than they solve. Thus Hall refers to a sensitivity analysis of results due to differences among models. This was conducted by three independent groups; he says "the main effect of (the experiment) is bewilderment" (347).

It would be useful at this point, to distinguish between the *sources* and the *sorts* of uncertainty. Classification by sources is normally done by experts in a field when they try to comprehend the uncertainties affecting their particular practice. But for a general study of uncertainty, we have to examine its sorts; as we call them, these are inexactness, unreliability and border with ignorance. Inexactness is the simplest sort of uncertainty; it is usually expressed by significant digits and error bars. Every set of data has a spread, which may be considered in some contexts as a tolerance or a random error in a calculated measurement. It is the kind of uncertainty that relates most directly to the stated quantity, and is most familiar to student of physics and even the general public.

A more complex sort of uncertainty relates to the level of confidence to be placed in a quantitative statement, usually represented by the confidence level (at say 95 % or 99 %). In practice, such judgements are quite diverse; thus estimates of safety and reliability may be given as "conservative by a factor of n". In risk analyses and futures scenarios estimates are qualified as "optimistic" or "pessimistic". In laboratory practice, the systematic error in physical quantities, as distinct from the random error or spread, is estimated on an historic basis. Thus it provides a kind of assessment to act as a qualifier on the number together with its spread.

Our knowledge of the behaviour of the data gives us the spread, and knowledge of the process gives us the assessment, but there is still something more. No process in the field or laboratory is completely known. Even physical constants may vary unpredictably. This is the realm of our ignorance: it includes all the different sorts of gaps in our knowledge not encompassed in the previous sorts of uncertainty. This ignorance may merely be of what is significant, such as when anomalies in experiments are discounted or neglected, or it may be deeper, as is appreciated retrospectively when revolutionary new advances are made. Thus, space-time and matter-energy were both beyond the bounds of physical imagination, and hence of scientific knowledge, before they were discovered. Can we say anything useful about that of which we are ignorant? It would seem by the very definition of ignorance that we cannot, but the boundless sea of ignorance has shores which we can stand on and map. Let us think of a border with ignorance as the last sort of uncertainty we can now control in practical scientific work. In this way we go beyond what statistics has provided in its mathematical approach to the management of uncertainty.

There is, of course, a rough correspondence between the "sources" of uncertainty derived from particular technical practices, and the "sorts" of uncertainty obtained by our previous analysis. This analysis provides a conceptual distinction among the technical, methodological and epistemological levels of uncertainty; these levels correspond to inexactness, unreliability and border with ignorance, respectively. There is no strict correspondence between these conceptual sorts of uncertainty and the sources derived from practice. To be sure, all data is affected by inexactness, and all computer models by ignorance. But as we have seen, ignorance may affects data, and inexactness occurs in the numerical analysis of computer models; thus the two classifications are distinct. A taxonomy based on sorts of uncertainty, like ours, enables the construction of a general notational scheme; and in it, the new and important category of "border with ignorance" can be operationalized.

Some reflective researchers perceive that the practical problems of uncertainty raise important conceptual issues. In an exhaustive survey on uncertainty in water quality models, M.B. Beck asks "why then, more specifically, has the analysis of uncertainty become so important, and what are the particular problems it poses?". His answer to this questions is:

Its importance is partly a reflection of the process of maturation typical of any subject of research. It is partly too a consequence of the liberating influence of the growth in the speed and capacity of digital computing equipment. The difficulties of mathematical modelling are not now questions of whether the equations can be solved and of the costs of solving them many times; nor are they essentially questions of whether prior theory... is potentially capable of describing the system's behaviour. The important questions are those of whether prior theory adequately matches observed behaviour and whether the predictions obtained from models are meaningful and useful. (Beck, 1987, 1393 – 1394.)

Beck's conclusions are arresting:

The implications, if not profound, are undoubtedly provocative. Consider the following conjecture. If the systems whose behaviour we attempt to describe are inherently imprecise, and if the observations that can be made of such systems are also imprecise, it is illogical to entertain algebraic or differential forms of equations as candidate descriptions of the system. (1433)

Describing what happens in practice when an expert is consulted on a water quality problem, he remarks:

This process is now explicitly subjective but then so too, implicitly (upon careful reflection) is the classical approach to model building. The distinctive roles of theory and observation have become blurred. It is not clear what has happened to the role of identification and diagnosis. Nor is it clear whether any principles of scientific method do or should govern this process (perhaps it has become "anarchistic" in Feyerabend's terms). (1433) (Feyerabend, 1975.)

Finally, he states:

The conclusion of this review is that a lack of model identifiability is unlikely to be overcome in the near future by improvements in the associated methods of numerical optimization. The profit to be derived from this failure is that model identification (of which parameter estimation is merely a part) should be more usefully viewed as a kind of forensic science, a painstaking piecing together and sifting of all the evidence obtained from a variety of lines of investigation, with the objective of providing a plausible and rigourous explanation of why the system behaved as observed. (1435–1436.)

Beck's study shows that in this field there are no formal structures whereby uncertainty can be eliminated or contained. If the process is "anarchistic", or (better) becomes a "forensic science", then uncertainty must be managed by skilled judgements, for the mathematical techniques will be irrelevant to this task. Hence, paradoxically, the powers achieved by computers for policy-related research increase rather than decrease the role of craft skills and judgements in the management of uncertainty.

M.C.G. Hall makes a similar point in connection with uncertainty analysis. Although it can never take into account a process that has been entirely omitted from a model... If used intelligently in conjunction with informal judgment and physical understanding, (it) can focus research on precisely those areas where omission of a process is likely to be important. (Hall, 1985, 362.)

Thus craft skills are complementary to mathematical methods in the effective management of uncertainty, even at the border with ignorance.

2.4. UNCERTAINTY-AVOIDANCE IN BUREAUCRACIES

When we move from the cognitive to the policy domain, the problems of uncertainty management change drastically. Actors in the policy process have their own problems of uncertainty that are conditioned by their agendas and by the style of their role. There will frequently be an unstable balance between two apparently contradictory approaches to uncertainty as it reflects policy: either to manipulate it to one's advantage in arguments, or to ignore or suppress it by administrative means. These situations arise because, in the context of decision, effectiveness in immediate practice outweighs abstract philosophical considerations about the True and the Good. As the legendary English judge said, on being criticized for his death-sentence on an innocent person, "I may sometimes be wrong, but I'm never uncertain".

There are strong reasons why bureaucracies in particular tend to prefer the avoidance of uncertainty rather than its manipulation. In connection with the "absolute" approach to risks based on precise numerical standards, H.S. Brown, lists the following reasons for their adoption:

- Once a standard is adopted, its application is simple and non-controversial.
- It is easy to justify and defend in court.
- It provides means of communication between all the technical and nontechnical participants of the risk management process on both sides of the issue.
- It *appears* (*sic.*) to be an objective process grounded in scientific analysis and free of value judgments.
- It relieves policy makers from a cumbersome burden of dealing with uncertainty and from being charged with imposing their own values and beliefs on society.
- It simplifies the problem by automatically determining the goals of risk management activities.
- It reflects a recurrent hope that we will find a scientific method for objectively resolving the problem of *How clean is clean*. (Brown, 1987, 235.)

An inevitable consequence of this approach is for standards to become "magic numbers" in the sense used by W.D. Ruckelshaus. As H.S. Brown says,

The common feature of absolute approaches is their search for universally accepted numbers, i.e. standards, guidelines, criteria. Once established, these numbers drive the cleanup process because they, in effect, define the term *Clean*. (234)

From a methodological point of view, it would seem more scientific for the agencies to adopt the "relative" approach, where "acceptable level is defined for each situation through the risk management process rather than used as an absolute goal for hazard management" (235). But to advocate this simplistically, would involve ignoring the realities of the American regulatory scene, where any premature admission of uncertainty is likely to be pounced upon by the other side in a ruthless adversarial contest.

In certain cases, however, the strategy of uncertainty-avoidance is question-ed even by agency spokepersons, particularly when a policy-number has been rejected or widely contested. Then they may demonstrate great skill in un-certainty management and thereby maintain their credibility. Thus in 1988, the United States Environmental Protection Agency (EPA) issued a "Radon Reference Manual", which included the policy-number of "4 picocuries of Radon per litre of air". This was said to represent the same risk of lung cancer as the (presumably acceptable) risk of ten cigarettes per day or two hundred (*sic.*) chest X-rays per year: a 5 % lifetime risk. When the number 4 was challenged, the EPA spokeperson remarked: "That's more risk than some people want to accept. We have to dispel the image that it is safe flying". This guideline was set as a level that could be achieved "fairly easily without frustrating, scaring or bankrupting homeowners"; the level was said to be not "safe" but "doable". The guideline was said, however, to have become "virtual gospel among homeowners, the real estate industry, and professional radon testers", thus functioning as a "magic number" in spite of official disclaimers (Purdy, 1988). The background to this story is that the EPA was originally instructed to set a "health-based" standard, which came to well below 4; and this was held up for a year by budget officials. The guideline was eventually set so as to have 90 % of homes deemed "safe", and also incidentally to save the Federal Government millions of dollars in decontamination costs for homes in the West that had been affected by Uranium tailings.

Bureaucracies can achieve quite high levels of sophistication in courting the public for that trust which is necessary for the management of uncertainty on their terms; The more enlightened are well aware that "if the public cannot evaluate the risk, they will evaluate the regulator" (Cantley, 1987). Samples of such effort in "risk communication guides", have been analyzed by H. Otway. Acknowledging that they avoid "numerical factoids that some techni-cal experts would have preferred", he argues that they are

essentially etiquette books for a how-to-do-it charm school. They have a strong tactical, problem-solving flavour, sub-dividing risk communication into *problems with the message, with the sender or source, with the channel, and with the receiver* (Covello, 1987).

Although ostensibly not intended to aid manipulation of the public (*the heart of effective communication is negotiation and coalition building, not manipulation,* (Covello, 1987)), they do incorporate implicit mental models of who is communicating what information to whom and why. They tend to assume that someone in authority, e.g., the plant manager, the regulatory agency, is communicating risk information to lay people 'out there', for whom the message must be formulated in language simple enough to be understood.

There is an identification with authority which takes for granted that those managing the risks are competent, honest and acting in the public interest. The context assumed is one of deciding about the acceptability of risk, albeit in an enlightened way that allows for public involvement. The impression given is that the main goal of communication is to explain risks and to provide reassurance that they are being responsibly managed. (Otway, 1988.)

2.5. CRITICISM: TECHNICAL, METHODOLOGICAL AND PHILOSOPHICAL

Sophistication in the management of uncertainty is only rarely taught. Most of those using statistics have been trained in straightforward application of techniques, and shielded from the awareness of methodological problems and pitfalls. Still less will they be instructed in the recognition of the manipulation of uncertainties in the bureaucratic or political context. There does exist a critical literature on statistics. The classic *How to Lie with Statistics* (Huff, 1954), first produced for trade-union negotiators in the United States, is still valuable for exposing the elementary techniques of the misleading use of numerical and graphical representation. Other books written by reflective staticians provide guidance for those specialists who are not content with the non-critical attitude of normal professional practice (see, for example, Feinstein, 1977, and Jaffe and Spirer, 1988). Also, important political criticisms of systematically biassed social statistics have been mounted (see, for example, Irvine *et al.*, 1979). However worthwhile, these different sorts of criticisms have focussed on statistics of one sort or another; they could well be supplemented by a coherent foundation for criticism and quality assurance for all quantitative information. This should include the phases of production, representation and function in a unified whole. It would need to be based on a clear philosophical understanding of the various complementarities, including objective and value elements, and knowledge and ignorance, manifested even in the simplest of quantitative information.

Such a philosophical understanding, combined with an historical perspective, can be used to explain why our present effort comes when it does, and also why its precursors have been so few, partial and isolated. For it is only within the past few decades that the traditional image of science, both as knowledge and as power, has been ripe for review and revision. Until quite recently, it would have been rank heresy to suggest that scientific knowledge is anything but unconditioned nuggets of truth, or that mathematics in whatever form could be erroneous or misleading. That ideology corresponds to the practical and political needs of science through the triumphant centuries of scientific progress from the seventeenth through the nineteenth (Pearson, 1892, 3). Only in the present century have the challenges for science been changing (and now at an accelerated pace), so that it is becoming a matter of common sense to see the essential task of science as coping with uncertainties rather than as rolling back the frontiers of ignorance. Before this present time, it would have been difficult for reflective individuals even to think through such arguments and conclusions, and the nearly impossible for them to gain any audience among scientists or the public.

It is in such ways that a prevailing metaphysics creates constraints both on individual imagination and creativity, and on social acceptance; and these are all the more effective for being unselfconsciously applied boundaries on common sense plausibility. Only when a new shared experience reveals the increasing inadequacies of an established world-view, does it become possible

for a society to begin the lengthy and painful task of philosophical reconstruction, always focussed on the most pressing problems of practice. In most
earlier periods, the problems presented to philosophy in this way were related
to religion and the social order. But now the environmental threats, resulting
directly or indirectly from the societal use of our sophisticated science and
technology, are displacing those traditional conflicts as crucial issues for
humanity. Hence a new philosophy of science, centered on the practical task
of the management of uncertainties in the policy-related sciences, is an appropriate response.

We are introducing our method for managing uncertainty and assuring
quality in the context of the policy issues that make it urgent. But devising our
notational scheme has required both the philosophical understanding and the
historical perspective which locate it in a broader context. In this way, the
notational scheme itself functions as the core of a coherent approach to the
management of uncertainty, and to the assurance of quality of information.
With this in mind, we can now give a preliminary description of the system,
which we call NUSAP (Funtowicz and Ravetz, 1987a).

2.6. THE NUSAP SCHEME, UNCERTAINTY AND QUALITY

The NUSAP scheme was designed as a robust system of notations for expressing and communicating uncertainties in quantitative information. The
essential principle of the system, according to B. Turner, is that "a single
number standing alone is misleading" (Gherardi and Turner, 1987, 11). **NU-
SAP** is an acronym for the categories **Numeral, Unit, Spread, Assessment** and
Pedigree. Going from left to right, we proceed from more quantitative to more
qualitative aspect of the information. The numeral entry may be a number, or
a set of elements and relations expressing magnitude. It may be in the form
of decimal digits, fractions, intervals (two-ended or one-ended), or ordinal
indexes sometimes expressed as verbal locutions (such as the familiar description of certain risks as "small" or "remote"; or even the soundings from
a Geiger counter as "click", "chatter", or "buzz").

The unit represents the base of the underlying operations expressed in the
numeral category (as, for example, grams, or $\$_{1989}$, or deaths per year). To
provide more power of expression to this category, we divide unit into a
standard and a *multiplier* (as in economics, $\pounds_{1989}.10^9$ or $\$_{1989}.10^{12}$ for national
accounts). The spread category normally conveys an indication on the inex-
actness of the information in the numeral and unit places. There are a variety
of expressions for spread, such as "$\pm n$", "$\pm p\%$", "with variance σ", "to
within a factor of n" or "over a logarithmic range of n". Assessment expresses
a more complex sort of uncertainty. It should express a judgement of the
(un)reliability associated with the quantitative information conveyed in the
previous categories. It may be represented through "confidence limits" and
"significance levels" of classic statistics; or alternatively through those of
Bayesian statistics. Alternatively less formalized expressions may be used in

contexts which do not fulfil the special conditions required for the formalized assessments. For example, one may wish to indicate verbally a qualitative scale of "total", "high", "medium", or "low" degree of confidence in a given numerical statement, or to describe the result of a risk analysis as "optimistic" or "pessimistic".

Pedigree, finally, is the most qualitative and complex of all the categories of the NUSAP notational scheme. Its role is to represent uncertainties that operate at a deeper level than those of the other categories. It conveys an evaluative account of the production process of the quantitative information. Quoting B. Turner, "is it (the number) based upon an exhaustive and detailed measurement process, or upon a snap judgement from someone over the telephone?" (Gherardi and Turner, 1987, 11). This category operationalizes the epistemological sort of uncertainty, border with ignorance, mentioned previously. It is represented in the notational scheme in a matrix form. This displays the degrees of strength of crucial components of the production process. Thereby it maps the state-of-the-art of the field in which the quantity is produced.

The NUSAP scheme also enables nuances and alternative interpretations to be clearly conveyed; and through its use, debates on the meaning and quality of crucial quantitative information will be conducted with greater clarity and coherence. It also protects against the abuse of information by the distortion of its meaning (usually by an unjustified precision of quantitative expression) after it has been taken up in the policy process. NUSAP could also prove to be effective in the normal practice of science, enabling researchers to evaluate more easily the materials they study and use.

By means of NUSAP, the more "qualitative" aspects of information can be invoked explicitly, for the effective evaluation of its quality. We note that the two senses of the term "quality" are involved here, the one meaning "non-quantitative" and the other "goodness". This serves as a reminder that even during the long period when the quantitative aspect of information has been taken to be the only genuine bearer of truth, the domain of the Good never came completely under the hegemony of numbers. We must also distinguish among the different senses of "goodness" itself, as these influence our use of "quality". There is first a sense of "goodness" associated with "production and function", as with things designed and made by craft-work, and intended for performing a prescribed function. There is also an aesthetic sense, as of those achievements and products which are classic and inimitable. And, finally, "goodness" has also its familiar ethical sense. The different senses of "goodness" may all apply to one instance; but, in general, there will be cases having "goodness" in one sense but not in the others.

By means of NUSAP, we can effectively relate uncertainty and quality. These are two distinct attributes, for information of low certainty may yet be of high quality for its function. We remarked on an extreme case of this in connection with deforestation in the Himalayas. We can base our distinction on terms defined in the British Standards for Quality Assurance. There the distinction is made between the "grade", denoting the general degree of elaboration and refinement of a product or a service, and its "quality" or fitness for purpose within that class. Thus "an hotel with few public rooms, no bars, no lift, etc., could be low on a grade scale, but its quality could be high if the limited services which it did provide were exemplary". On the other hand, "a high grade article can be of low quality, if it does not meet its specification" (British Standard Institution; 1979, 9). In our case, the "grade" is a gauge of the degree of inherent uncertainty in the information, as express-ed in its pedigree code. For example, there is no point in ecological modelling trying to emulate experimental physics in its control of uncertainty. However, *within* a given grade, characteristics of a field of practice, information may be of greater or less quality, depending how well the uncertainty is managed, and how well it fits its purpose.

Assessments of quality and their use in quality assurance, have recently been appreciated as essential to successful practice in industrial production; the Japanese example is well known to us all. Conversely, we now know that major industrial disasters are caused, not by chance or by God, but by lapses in quality-control. We are less familiar with the idea of quality assurance in scientific information; yet it is equally fundamental. In the received philoso-phical view, science is seen as 'knowing-that", to the exclusion of "knowing-how". It is believed to be about the eventual attainment of truth. The cate-gories of quality, and of controlled uncertainty, appropriate to "knowing-how", have no place in the traditional philosophies of science. These have been concerned with normative, idealized reconstructions of science rather than starting from the practice which has made science a model for successful human knowledge.

In spite of its being ignored by philosophers, quality assurance is an integral, ubiquitous aspect of the social activity of science. Although no philosopher has discussed peer-review of refereeing (leaving them to such lower orders as sociologists), these are the stuff of the governing of science. It is in such activities of control that judgements of quality become explicit. Normally they are informal, tacit, sometimes even unselfconscious communal knowledge, whereby research skills are transmitted and maintained. When we think of scientific knowledge as produced by craft skills which are shared and presup-posed in an expert community, the categories "true/false" and "verifi-cation/falsification" reveal the severe limitations of their explanatory power. To be sure, good scientific work has a product, which should be intended by its makers to correspond to Nature as closely as possible, and also to be public knowledge. But the working judgements on the product are of its quality, and not of its logical truth. The criteria of quality may be highly specific to the field, and will normally relate more to the production process than to tests on the product. There are, of course, clear analogies with industrial production; but there is an important difference, in the complexity and specificity of the operations and skills for each research field and even project. The closer analogy would be with prototypes and pilot plant operations, or innovative technological systems with nearly unique copies.

In scientific work, uncertainty can never be banished, or even controlled by a routine. High quality of work depends on the skilled management of uncertainty. This principle should not be strange to working scientists, for whenever a statistical procedure or test is a applied, uncertainty is being managed by technical means. All these techniques incorporate value, as in error-costs of false positives and false negatives. To be aware of these aspects of the apparently formal manipulations of statistics distinguishes the true craftsman from the unskilled scientific operative.

Since we are considering information as an input for decision processes, the analogy with physical artefacts, either components or tools, can be illuminating. For them, quality is assessed through a variety of categories. One is more quantitative, a tolerance corresponding to our spread category. This may be in the precision of the machining of a component, on the tightness of performance to given specifications. But more qualitative categories enter, including the fineness of materials, and the history of the make, the model, and perhaps of the copy itself. These latter elements constitute a pedigree; and could doubtless be formalized like the one for NUSAP. The overall quality assessment is then derived for that item from its spread and pedigree, as an estimate of reliability in use (perhaps balanced against cost) in relation to its production and function. For a simple example, a common nut-and-bolt combination made of cast steel will have different criteria of quality from one made of machined brass; different functions call for appropriate criteria of quality, with their own sorts of production and sorts of costs.

We have seen that uses of quantitative information include some for which there is no simple analogue in science or even in industry. These include the many sorts of "rhetorical" uses that occur in policy and political debates. In such contexts "quality" and "reliability", as evaluated by particular actors, may relate mainly to their own value commitments and personal age. While these are "objective" in some important senses, and not to be dismissed as "purely subjective", still they are numerous in type, and complex and sometimes contradictory in character. To try to define "quality" and "reliability" in relation to each one of this whole set of uses, is a task whose limits cannot be discerned. For some particular interactions, the assessment of quality and reliability is possible, and we provide examples of this later on.

Those who become skilled in using NUSAP will know how to apply it, with prudence and discretion, to new situations of policy or political debate. They will become aware of the limits of plausibility and fruitfulness for formalizations of real experience in practical cases. Finally they will be also sensitive to the possibilities of manipulation of information, and indeed, of NUSAP itself; and will be able to react appropriately.

2.7. NUSAP: PHILOSOPHY AND PRACTICE

The NUSAP scheme, although originally designed for the management of uncertainty and for quality assurance in quantitative information, also has implications for the philosophy of science. That field is now experiencing a multiple crisis. Philosophers recognise the irrevocable collapse of the programme to develop logical structures explaining the necessarily successful practice of natural science (see, for example, Shapere, 1986). This failure in epistemology is paralleled by a similar one in the methodology of research, particularly in the policy-related fields. Both of them stem from the dominant traditions in philosophy; these have exalted the "knowing-that", which was to be kept uncontaminated by contact with "knowing-how". A philosophy of science that abstracts from practice and its uncertainties must eventually become recognized as irrelevant and sterile.

The two problems of failure actually come together, implicitly at least, on those issues where in one way or another the traditional methods of science have revealed their inadequacy. In debates on large scale problems, such as industrial projects constituting "major hazards" or major environmental intrusions, or in the speculative technologies of nuclear armaments, uncertainty is politicized, and the dividing line between science, nonsense and fantasy can become very difficult to discern (Ravetz, 1989). The traditional methodologies of scientific research offer insufficient protection against the corruptions of reason which are encouraged in modern conditions, even in our dealings with the world of Nature. We need a methodology which both comprehends the issue of quality of information, and also operationalizes the degrees of quality, from the best to the worst. Otherwise we have at best only vague pronouncements, to provide an alternative to the old faith in science as embodying the True and the Good. In this way the practical uses of the NUSAP system are of direct philosophical importance.

The NUSAP scheme may also contribute to an enrichment of our inherited common sense and knowledge about natural phenomena. It may provide a basis for transcending the seventeenth century metaphysics in which geometrical reasoning was to supplant human judgement as the route to real knowledge. Instead of erecting some general, all-encompassing, polar-opposite alternative to our dominant "reductionist" science, be it in the form of a "holistic", "romantic", "idealist" or "voluntarist" philosophy, we can in a practical way exhibit the essential *complementarity* of the more quantifying with the more qualifying aspects of any quantitative statement. Human judgements need not be viewed as inhabiting a separate realm from exact mathematical statements, bearing a relation that is either hostile, mysterious, or nonexistent; but rather as a natural and essential complement to the more impersonal and abstract assertions embodied in a numerical expression. When this dialectical insight, made familiar in everyday experience, is available for philosophical reflection, than we may be in a position to go beyond Galileo's fateful pronouncement that the conclusions of natural science are true and necessary and that "l'arbitrio humano" has nothing to do with them (Galilei, 1632, 53). (For an application of the ideas of complementarity to technology, see Pacey, 1983.)

If we were to make a simple opposition of "knowing-how" to "knowing-that", and call for a replacement of one by the other, we would be committing the same philosophical error as is traditional in the philosophies we criticize. When we emphasize "knowing-how" and the craft, informal, partly tacit element of science, we are correcting an imbalance in the appreciation of two complementary, interpenetrating aspects of that practice. To value "knowing-how" to the exclusion of "knowing-that" would quickly lead to superficiality and banality in practice and results. Our philosophy, on this issue as on others, promotes a dialectical synthesis.

A new epistemology which includes "knowing-how" through the concept of quality, should enable us to reinterpret received views in its terms. All philosophers of science have made recommendations for good practice, that is for high-quality work. Thus Popper advised scientists to advance the most bold and general hypotheses for attempted falsification. The same holds for Kuhn, in his ambiguous, ironic fashion, as seen in his term "mature" and perhaps also "normal". Feyerabend tried to show that scientific and moral quality were equally irrelevant to the successes of the great masters, notably Galileo.

Using the elements of craftsmanship and quality, in relations to the "knowing-how" aspect of science, we can reformulate the classical demarcation problem between science and its spurious imitations This problem was at the heart of the programme of traditional philosophy of science, through all the twentieth century and before. Quite simply, we may ask, does a field of expertise support the work of quality assurance? If so, it is genuine; if not, not. Answering such a question is not easy, since any sham-science will have sham-institutions for the pretense of quality assurance. Nor is the demarcation in practice a simple black-and-white distinction. In fact, we can reformulate this demarcation principle to be a usable criterion for quality of areas of scientific expertise. We scrutinize the actual systems of quality assurance, and on that basis we make our assessment. This examination does not require the esoteric technical knowledge involved in evaluating research results; it can be accomplished by attention to group behaviour, especially openness to criticism. (Here the Popper of *The Open Society*, as opposed to Popper the epistemologist, receives his due credit.)

The NUSAP scheme can thus become a tool with many uses at different levels, for the effective expression of quantitative information. But like any tool it has its own inherent limitations. Most important, NUSAP, like any other representation, cannot enhance the reliability of information beyond that of its source. Also, since it is designed for flexibility, it cannot be applied automatically or unthinkingly; and it presupposes an awareness that all quantitative information has its qualifying aspects, corresponding to its various sorts of uncertainty. Applied without such an awareness, or in a meretricious way, it could be misused or abused; but no intellectual production is immune from such defects. At the very least it brings certain aspects of information which were previously obscure and neglected to focal, disciplined awareness. We might say that even if a debate involving NUSAP does become corrupted, at least it takes place at a higher level of sophistication and awareness than previously.

Within such limitations NUSAP can help to make scientific information more effective, in several ways. First, simply by displaying the qualifying aspects of the information, it provides guidance for its skilled and prudent use. Through the listing of several alternative representations for the same information, the system can provide a highly articulated and nuanced expression in a convenient and compact form. The system also provides guidelines for the inquiry and elicitation involved in the evaluation of its quality. By such means, it makes possible a quality assurance of information. This is no less necessary in the area of policy-related research than in industry and administration. The use an analogy from science, we may say that NUSAP cannot be an alchemical process that transmutes base material into noble; but like good chemistry it can refine that material for maximum quality and effectiveness.

CHAPTER 3

THE MATHEMATICAL LANGUAGE

The new requirements on scientific information used as an input to policy decisions have revealed inadequacies in the notational systems used for the expression and manipulation of quantities. The new tasks of quality-control and communication are not well served by our inherited notations, nor by the associated widespread belief in their clarity and exactness. But we should not think that a natural and faultless numerical system has suddenly been stretched beyond its limits on applicability. We shall see that there are significant problems in the ordinary use of our arithmetical system. These derive partly from the historical development of numbers as used for the exact counting of small collections of discrete objects, and not for estimation. Also, a philosophical analysis shows that arithmetical language, in spite of its formal appearances, shares some significant features of ordinary, informal language. These include the importance of context in the determination of meaning, and because of its inherent ambiguity and vagueness, the need for judgement in applying the linguistic system to practice.

Such problems could be neglected so long as craft skills sufficed for the management of uncertainty and for the assurance of quality in mathematical operations. In our present situation, where highly uncertain quantitative information is used so extensively in policy-related research, the conditions are ripe for a combination of historical perspective with philosophical understanding, for the creation of an enriched arithmetical language appropriate to the new requirements.

3.1. HISTORICAL PERSPECTIVE

Historically, numbers developed by successive abstractions from the practical operations of counting. The earliest records of numbers show them as adjectives, always appearing as modifiers on some particular sort of object or collection. Only later did they become independent, so that the sum "one soldier plus one soldier makes two soldiers" could be seen as a special case of "one plus one equals two". Only then could there be calculations with symbols of abstract quantity. Thereafter, the whole edifice of mathematics as we know it could be constructed, becoming an independent conceptual system of ideal objects. These presented new opportunities and challenges. On the one hand, there emerged "pure mathematics", the discovery of new ideal entities and proof of new properties and structures among them. On the other hand the foundational difficulties were encountered almost immediately, as some objects displayed counterintuitive or paradoxical properties (first the quantities that were "irrational", and then pairs of lines that were "parallel").

In both of these ways, mathematical thinking, particularly in Western culture, tended to become alienated from ordinary practice. This can be seen historically as the price mathematics has paid for its great successes, in its own development, in its application to science and technology, and in its influence on society and culture.

Of course, there has always been an interaction between the theoretical and practical aspects of mathematics. As in the case of geometry, practice has supplied some basic problems and techniques; and abstract systems when articulated have contributed to more developed, universal and powerful theories, as for example "the calculus" applied to science and engineering. But for many reasons, both technical and socio-cultural, these two streams of mathematics have never been brought into harmony. The gap between them is only occasionally appreciated; the usual presentation of all mathematics as essentially exact clear prevents an understanding of how these intellectual system relate to practice. Because of this the role of skills and judgements in the application of mathematics is generally ignored.

The way in which abstract mathematical systems may help or hinder practice will depend on historical circumstances. In the nineteenth century, "practical arithmetic" was widely taught in the schools. But in the present century, the rise of academic research in pure mathematics has produced a tendency to re-cast mathematical education at the lower levels into directions derived from, and serving, the research frontier. This fashion, combined with a faith in abstraction and axiomatics as the essence of mathematics, was exemplified in the classic work of the Bourbaki school (see, for example, Dieudonne, 1970). The result was the "new mathematics" of the sixties and seventies. This was entirely devoted to the teaching of the elite skills of manipulating abstract structures to all schoolchildren. These did not complement practical skills, but effectively alienated even simple arithmetical operations from experience. Thus,

The formalist style gradually penetrated downward into undergraduate mathematics teaching and, finally, in the name of the new math, even invaded kindergarten, with preschool texts of set theory. A game of formal logic called WFF and Proof was invented to teach grade-school children how to recognize a well-formed formula (WFF) according to formal logic. (Davis and Hersh, 1981, 344.)

This pedagogy, based on disapproval of rote-learned practical craft skills, had its analogue in the "global methods" for teaching reading while ignoring the alphabet. As a result of such innovations, for some years children emerged from the best schools unable to spell or to do arithmetic (Kline, 1974).

Now the situation has changed, and is less coherent and more open. The "new mathematics" passed through its cycle of increase and decline in plausibility, analogous to the stylistically similar "new architecture" of Corbusier and the Bauhaus. Philosophical critiques of the influence of positivism in mathematics were advanced by Lakatos, with his dialectical, quasi-practical approach (Lakatos, 1976); and there has been renewed interest in intuitionism (see, for example, Dummet, 1977). The collapse of positivism has helped to weaken the ideological appeal of formalism, which in any event had long since lost its philosophical rationale with Gödel's theorem (1931), or even earlier (see, for example, van Heijenoort, 1977). At the same time, computers have been coming in ever more strongly, providing job opportunities and new criteria of interest and even of proof. Most notable in this direction was the computer based "proof" of the classic "Four-Colour Theorem" (Appel and Haken, 1978). This was accomplished by means of an exhaustive survey of a multitude of special cases, thus introducing a quasi-empirical, inductive style of proof, in contrast to the universal, ideal and deductive style, which was traditional in pure mathematics.

Teaching is left without a unifying conception of the essence of mathematics; it is then increasingly responsive to the new demands of market applications and jobs which are dominated by computers. Now mathematics is in a condition analogous to that of architecture and other arts of creative design. The age of a hegemonic style based on an abstractionist aesthetic, has passed. In mathematics, foundational studies are no longer tied to a positivistic philosophical programme. In fields related to the natural sciences, the traditional assumptions of Laplacian determinism (which had been left untouched outside the "quantum" domain) are complemented by the mathematical study of phenomena involving "chaos". Such studies, using computers as an essential tool, have also produced enchanting works of art in the display of the intricate mysteries of the properties of simple mathematical systems (see, for example, Gleik, 1988).

All these developments provide a perspective on why our proposed innovations are now both necessary and practical. The faith in formalized mathematics as the embodiment of truth and rationality, so strong in the Western civilization since Pythagoras, is now at a low ebb (Kline, 1980). It becomes possible to articulate and to gain support for a sort of mathematical thinking that moves on from abstractionism. This is based on the dialectical relation between theory and practice, incorporating skills and judgements into conceptual structures, and developed in close association with a methodology for the quality assurance and communication of scientific information.

3.2. MATHEMATICAL LANGUAGE AND UNCERTAINTY

The methods and results of quantitative science are communicated using a mathematical language. The symbolic and formal appearance of such language has led to a widespread belief that it is a part of logic. But the symbolism of mathematics should be seen more as a fertile abbreviation of prose than as a logically structured language with its rigid rules of syntax and transformation. The mathematical language of science is more similar to "natural" languages, whose terms stand for concepts deriving from practical experience. They share all their open texture, including vagueness, ambiguity and even contradictions. Paradoxically, these properties (including contradiction) are a source of the fruitfulness of this language (arithmetic included); we shall see how "zero" exemplifies these features. The fact that the rules of ordinary arithmetic appear to admit of no exceptions has led to the common ignorance of this feature of mathematics. But as we will soon show, the simplest of arithmetical operations, when applied to particular practical cases, require a re-interpretation of the symbols, and the invoking of higher-level rules, lest nonsense result.

The language of mathematics is a hybrid produced by a long historical development. It combines an abbreviated language derived from practice with explicit rules of formation and transformation relating to the logic of valid inferences. For this language to be applied effectively there must also be rules of translation, so that any given set of objects in the empirical world can be matched to its analogue in the mathematical ideal system. These two sorts of rules could be seen as corresponding to the semiotic categories of syntax and semantics, respectively. The third category in a semiotic process, namely pragmatics, involves the user of the mathematical language and has never been formalized to the same degree. For "use" involves grappling with particular cases, as well as invoking intentionality and values; and hence tacit judgements and craft skills are necessarily involved. In every practical case, the set of explicit rules for use will be insufficient for the proper operation of a system in all contingencies (judgements must be applied, for example, for over-riding the rules). The standard mathematical language by its very form conceals uncertainty and has the effect of inhibiting even the awareness of uncertainty; it inevitably imposes a particular view of reality and practice on its users.

The NUSAP notational scheme, designed for the communication of quantitative scientific information, exhibits uncertainty in all its forms. It guides judgements and fosters skills in the management of uncertainty. It helps accordingly to enrich the conceptions of reality and practice, of both producers and users of information. NUSAP is descriptive and taxonomic rather than analytical and formal. We do not intend to provide a new "calculus" whereby symbolic manipulations can carry all arguments about "uncertainty". We should explain why we are departing from the mainstream approach in this field, which has always attempted to control uncertainty through the development of appropriate formalisms.

Of course, technical uncertainty (as represented by error bars), and in some cases methodological uncertainty (as in confidence levels) can indeed be expressed through a calculus. Similarly, "expert opinion" is formalized by Bayesian statistics, and vagueness of class-membership is described by fuzzy set theory. Such formal techniques are effective tools for the management of those sorts of uncertainty, and could usefully be complemented by the NUSAP approach. But the more complex aspects of methodological uncertainty, and epistemological uncertainty as well, are outside the boundaries of any calculi. This is why, rather than attempting to formalize the NUSAP scheme up to the limits of plausibility, we have made the design decision to keep it simple, robust and perspicuous.

3.3. FORMALIZATION AND INFINITE REGRESS

We can justify our decision by showing that a complete formalization of uncertainty is impossible. Two examples (from Bayesian statistics and fuzzy sets) will illustrate this point, and also show the way into the conceptual argument. In connection with Bayesian probability distributions provided by experts, M. Henrion discusses the issues raised by the use of "second order" probability distributions. Along with the original uncertainty being estimated, there is a higher-level uncertainty, that associated with the estimates of uncertainty themselves. For example, the application of Bayesian methods does not completely eliminate uncertainty; there is a residual uncertainty, for the experts' assessments are not perfect knowledge. Thus, we are in presence of a sort of "uncertainty^2"; what is to be done about this? According to Henrion,

A more sophisticated approach to representing uncertainty about probabilities is to use a probability distribution over the probability distribution, that is a second order probability distribution. Questions have been raised as to what exactly these might mean, given that probabilities are only defined for well-specified quantities. Assigning a subjective probability distribution to a frequency (objective probability), such as the failure rate in a large population of similar components of a nuclear power plant, poses no theoretical problems. But it does not appear that a subjective probability itself can pass the clarity test, since it is one person's opinion and not open to empirical measurement even in principle.

One useful interpretation of your probability distribution on a probability is as your current opinion on what your posterior probability would be after you have observed some relevant information... The more informative the source, the greater the uncertainty of this distribution, since the more you are likely to change your opinion. (Henrion, 1988, 14.)

Thus there is a clear appreciation of the issues involved in second order estimates but still undeterred some seem to be prepared to iterate the Bayesian procedure. Regardless of the questions of interpretation discussed by Henrion, there is a systematic problem appearing, once one has embarked on estimating uncertainties of uncertainties. For clearly, this is not a way of eliminating the residual uncertainty; it will be there after the second estimation as well. Thus the attempt to solve the problem of uncertainty *within* the framework of formalisms, leads directly to an infinite regress.

This phenomenon of infinite regress was explicitly mentioned by Goguen in connection with fuzzy set theory. He states the problem as follows,

Is it a paradox that the degree of membership used to indicate a degree of uncertainty is itself very precisely given as a real number? More generally, is it a paradox that this theory of imprecision is very precise, and even based on ordinary crisp mathematics? (Goguen, 1979, 49.)

His solution in his earlier work was to iterate, obtaining "fuzzy fuzzy fuzzy sets" or "type 3 fuzzy sets", and in general "(fuzzy)^n sets". He then takes "the limit, as $n \to \infty$, of (fuzzy)^n sets, provided that one restricts consideration to fuzzy sets (at each level) which are continuous", and remarks:

However, the basic point here is that one cannot escape the difficulties of obtaining and justifying truth values simply by running up the logical type hierarchy (even if one takes it to the limit and goes over the top) (*sic.*). In fact, since one actually gets more values which have to be dealt with, the problems becomes intensified. (58)

He concludes with a "radically pessimistic note",

There seems to be little room for belief that fuzzy set theory... will provide an ultimate answer to mankind's problems and uncertainties about the future of large systems, either in general, or in particular important instances. (65)

The problem of recursion and infinite regress as raised by uncertainty has an analogue in the field of proof-theory (or "metamathematics"). Here, it is assumed that the language being studied is purely formal; its symbols and transformation procedures are considered syntactically and not semantically (e.g. independently of any possible meaning). In its early period, this field was developed as part of David Hilbert's programme for the complete formalization of mathematics. The results of proof-theory and the practical experience of mathematicians eventually dictated otherwise. The unavoidable necessity for informal languages manifested itself in two ways. First, discussion of any formal language cannot be conducted within the language itself, but requires a "metalanguage" (and if *that* is to be studied rather than merely used, we need a "metametalanguage", and so on). Hence, if the theory of the original language ("object language") is to be something other than a very dense set of marks on paper, the recursive sequence of formal languages must eventually stop somewhere, and ordinary language must be used. Further, the fact that metamathematics is done by and for humans was decisive. Thus,

The metatheory belongs to intuitive and informal mathematics... The metatheory will be expressed in ordinary language, with mathematical symbols,..., introduced according to need. The assertions of the metatheory must be understood. The deductions must carry conviction. They must proceed by intuitive inferences, and not, as the deductions in the formal theory, by application of stated rules. Rules have been stated to formalize the object theory, but now we must understand without rules how those rules work. An intuitive mathematics is necessary even to define the formal mathematics.

We shall understand this to mean that the ultimate appeal to justify a metamathematical inference must be to the meaning and evidence rather than to any set of conventional rules (Kleene, 1964, 62).

Kleene's argument can also be stated in terms of recursion of rules as well as of languages (or theories). The philosopher L. Wittgenstein discussed the ambiguity of rules as directions for practice, and introduced the possibility of an infinite regress of "rules to interpret rules" (Kripke, 1982, 83). In this instance it is the sequence of rules-about-rules that must end in informality. In both these cases, languages and rules, the decisions on when to end the recursive process are necessarily governed by judgements.

3.4. RULES: WHEN TO OVER-RIDE?

These conceptual discussions parallel practical experience of many sorts, ranging from domestic electronic equipment to nuclear power stations and air traffic control. In every case the set of explicit rules can be insufficient for the proper operation of the system. Judgements must be made for interpreting or occasionally even over-riding the rules. These depend on the skills of recognition of the degree to which a given situation matches that in a rule-book, and also the skills of operating in circumstances not envisaged in the rule-book. In the absence of such skills, either blind obedience or random panic reactions are the only alternatives. A simple example is the apparently straightforward exercise of following an instruction manual, for setting up some home electronic equipment. The manual's formal instructions, though apparently followed in detail, frequently do not suffice for getting the system to work. The user is then thrown back on his or her resources: do we simply repeat the operations yet again, or do we add something extra, trying some combination or sequence not mentioned in the manual? Such decisions on over-riding the formal instructions can become urgent, and quite fraught for the individuals concerned.

The first industrial example, nuclear power stations, exhibits the complex interface between standard operating procedures and the over-riding judgements to be exercised in emergencies. The most obvious case is the point at which operatives should recognize that a situation is not any of those described in the rule-book; it then requires a decision to over-ride the automatic systems by manual means. Thus there is an official recognition of the incompleteness in practice of any rule-defined system. In addition, there is no sharp or clear boundary between the cases "in the rule-book" and those not. For, to establish a sharp or clear boundary would require another rule-book – should we say a "metarule-book" – which is again part of a recursive sequence that must end in informal judgements by a skilled operator. The problem of the over-riding of rules appears as a sort of inverse to the problem of indefinite recursion. In this instance, there is no availability of an unending sequence of metarules; and so when skilled operators encounters a situation which is not certain to be covered by a rule, they must use their judgement to decide.

The other industrial example, air traffic control, shows how the smooth operation of a system may routinely require the exercise of judgements for overriding the formal rules. Air traffic controllers have various ways of "working with the system to work around the system". Among these is "stack jumping" where an outbound aircraft is directed to climb over an inbound one, a procedure not recognized by the system. In this way, traffic can be kept moving, in the traffic controller's judgement, safety is not compromised (Hughes, 1989). On their part, the pilots also contribute to beating the system, particularly when computer control becomes onerous. Thus it has been reported that "when pilots want to begin their descent earlier than the point preprogrammed into the computer" (perhaps on following changed instructions from air traffic control) "they have to trick the electronics by putting in false information about tail winds." (Gavaghan, 1989.)

We notice here that skills are involved in over-riding the formal rules; indeed we may say that a large component of the craft skill on any practical job consists in doing just that. Formal rules may well be framed in the expectation that they will be got around; their function is the for general guidance, as well as for cover against liability in compensation cases. We may say that disasters occur when operators are forced by various pressures, usually informal, to evade or to break the rules beyond the point where their skills and judgements can ensure safety in the event of things going wrong. The official explanation of "operator error" then amounts to the classic tactic of "blaming the victim". Also, the social history of technological innovation can be seen as containing a strong element of development for the sake of de-skilling operations. This reflects the social division of the enterprise, the operatives pitting their (sometimes secret) "knowing-how" against the formal "knowing-that" of management and their hired experts. In the tradition of Taylorism, some experts still believe that all operations should be explicitly defined and programmed. Fortunately, this is a fantasy, which computers have not changed, as the real world will always defy total specification of tasks or of any other sort of knowledge; and formal rules will always need to be over-ridden for any job to get done.

There is a natural connection between the cases of infinite regress in the formal systems for describing uncertainty, and over-riding of rules in the practical situations. For no real system can iterate to infinity; and if a formalized system has no instructions for a user to leave it, then a resort to personal judgement amounts to an abandonment of the system and, in effect, an over-riding of its rules. In well-designed systems of monitoring and control, the transition from explicit rules to private judgements will not be abrupt. There may be a few metarules in the form of general guides or hints rather than directives; and then there will also be procedures for reporting and consulting with colleagues, superiors and experts elsewhere. In such ways, the transition is buffered, by steps which progressively decrease the formalized contents and increase the elements of personal judgement.

In a similar way, the NUSAP notational scheme provides a set of places for qualifying information which is decreasingly "hard" and increasingly "soft". The categories numeral and unit will be expressed in general as they would in ordinary scientific practice; the same with spread, although with NUSAP we can allow for its essential vagueness. The assessment category can be represented by indices of quite general form, or even by words describing strength or quality. Finally, the pedigree place will be occupied by a summary of a matrix embodying a taxonomy based on history and experience. Because of this polar structure, NUSAP is not susceptible to the same paradoxes and traps as the purely formalized approaches. Through the pedigree entry of a notation, the user is led on to the informal judgements of reliability, quality and available knowledge, which in ordinary life guide the partial and provisional acceptance of assertions.

One may indeed ask about the pedigree of a pedigree or (pedigree)^2. The answer to this question will depend on the exact meaning of the question; if a large set of pedigree entries are in question, then the circumstances of their production process are a legitimate topic for enquiry, and if appropriate, may be cast in matrix form. But if one is seeking for the warrant for a particular pedigree matrix, or even for the concept of pedigree itself, then an informal answer is more suitable. In other fields where reliability of statements or evidence must be attested (as in jurisprudence), a mixture of more formal and less formal elements is deployed. With NUSAP, we attempt the same for policy-related research. Our design decision is to forego the benefits of highly articulated calculation with uncertainties in the system as a whole (although the various calculi and their results can be incorporated, as in spread), for the sake of robustness, transparency and applicability; and these properties are reflected in the freedom from paradoxes such as infinite regress.

3.5. AMBIGUITY AND VAGUENESS

The cases of uncertainty in following rules can be seen as involving ambiguity and vagueness. Ambiguity occurs where there is a discrete set of possible relevant meanings, and uncertainty about which is the appropriate in an instance of use. Vagueness refers to a continuous spectrum of interpretations, either because of the absence of precise demarcations (as in polar pairs of words in ordinary language like hot-cold), or because of a multiplicity of criteria of use, leading to a set of overlapping meanings (as, for example, in ordinary language, the word "game". See Hospers, 1967). For reliable technical practice, it is necessary to reduce ambiguity and vagueness as much as possible; but it is an illusion (sometimes dangerous) to believe that they can be totally eliminated by means of new explicit rules or definitions.

For an example of ambiguity, we take the case of faulty monitoring systems; these can involve "false alarms"; where the fault may lie in the process itself or in the monitoring system. A red light in a cockpit display signals fire in a particular one of the engines of the aircraft, but it may also be due to a fault in the warning system. A monitoring system whose ambiguity is familiar to operatives can actually increase the hazard, as they may then assume that every danger signal is a false alarm. The attempt to eliminate ambiguity by the introduction of new back up monitoring systems is analogous to that of the "metarule-books"; unless done with great skill, it can merely multiply the ambiguity of the signals, and the confusion and danger.

As an illustration of vagueness, we have the "border-zone phenomenon". Suppose that an operator is instructed to make an entry in a log-book of incidents every time a significant indicator passes a threshold value. There will always be instances where the value is *nearly* there, but not quite; whether an entry is made will then depend very strongly on the system of rewards and sanctions influencing the decision to record. To make the threshold a zone merely gives the operator greater latitude in interpretation. The statistics on such incidents will, in spite of their precise form, be affected by operators' interpretations and judgements in the borderline, vague instances.

These examples of ambiguity and vagueness show the impossibility of eliminating judgements from the operation of complex technological systems. We can now extend the argument to those very familiar, apparently trivial intellectual systems, numbers. Just as in more sophisticated cases, the user of numerical systems must be ready to make judgements for the applicability of standard rules, and to recognize vagueness and ambiguity in formally-expressed concepts. Normally this is done unselfconsciously, as there is little recognition of these phenomena, and no formal teaching about them. With the NUSAP system, we can make explicit all these forms of uncertainty; and because of its partially formalized design, we can manage them without falling into infinite regress.

3.6. ARITHMETICAL RULES: THE FOSSILS JOKE

A few simple examples will illustrate how ordinary arithmetical language and rules of operation can be misleading. By their appearance they seem to guarantee perfect exactness of expression; and indeed this is one of the motives for their being accepted as the standard for science. But there are two reasons why this impression is wrong: first, that the inevitable inexactness is hardly ever conveyed within a quantitative statement; and second that the notation itself contains deep ambiguities, which are fertile for practice but (when unnoticed) are also a source of confusion. Schoolchildren are told that arithmetic is simple in all respects, but then encounter puzzling and paradoxical properties of these standard mathematical tools, for which they have no preparation or guidance.

A joke about fossils will illustrate many of the points we are making. It relates to a museum attendant, who was heard telling schoolchildren that a particular dinosaur bone was fifty million and twelve years old. Asked how he knew so precisely, he answered that it was quite easy: when he came to work there, the fossil was labelled as 50,000,000 years old, and that was twelve years ago. Clearly, the attendant was somewhat simple-minded; but what was wrong with his reasoning? He did the sum 50,000,000 + 12; and as he has learned at school following the arithmetical rules of adding place by place, he obtained the precise result 50,000,012. Should he have *not* used standard arithmetic?

What he lacked here was judgement, about the interpretation of the legend 50,000,000. Clearly, the string of symbols did *not* mean that a large number of years had been counted and added exactly to 50,000,000, a number lying between 49,999,999 and 50,000,001. The six last zeroes taken together stand for "million", the effective unit of counting in this case. This should have been appreciated by the attendant. Then he would have seen that the sum 50,000,000 + 12 should have been added by an artefactual arithmetic, giving the answer 50,000,000. Thus, the most elementary of mathematical operations, when correctly applied in such a context, has a surprising similarity to Cantor's cardinal arithmetic of transfinites, where, for example, "$Aleph_0$" (the lowest order of infinity, corresponding to the set of integers) when added to n (any integer) gives the sum $Aleph_0$ itself. Suddenly we see that an ordinary mathematical operation, that every person is taught at school, must be replaced on some common occasions by an artefactual arithmetic, similar to one known explicitly by very few. Yet the awareness of the inappropriateness of the rules of ordinary arithmetic in certain contexts must be widely understood, otherwise the story would not be a joke. (The point would be unchanged even if we wrote 5×10^6; for the standard rules of arithmetic are not automatically suspended for the "more scientific" exponential notation.)

When to over-ride the rules of the ordinary arithmetic, and apply those of order-of-magnitude arithmetic is a matter of a correct judgement of the context in which the operation is carried out. This is taught nowhere, and yet (since the story can function as a joke) is assumed to be known by every numerate person. This is a rather strange kind of knowledge; what is not explicitly taught is assumed to be common-sense, while what is explicitly taught is the belief of fools. And this is not about life-skills of adults, but the subject-matter of one of the most basic subjects in the school curriculum. It appears that arithmetic, or more generally, the practical interpretation and use of quantitative expressions, requires a special kind of common-sense, which knows when it is appropriate to over-ride the formal rules learned at school.

3.7. ZERO: COUNTER OR FILLER?

The "fossils joke" has shown how competent practice includes the knowledge of when to over-ride standard rules. An examination of the sum used for the example, illustrates the ambiguity of a very basic symbolic notation. Like any natural language, our mathematical language possesses ambiguities that are fertile, indeed essential for its operation. Here we focus on the "zero", a strange quasi-number. It was introduced quite late in human cultural history, around the sixth century A.D. probably in Indo-China. It can be interpreted as an operationalization of the Hindu concept of the Void in the context of the (then incomplete) Chinese place-value numerical system (Needham, 1956). It is known to have a non-standard arithmetic, where $0 \times n = 0$, and $n/0$ and $0/0$ are meaningless. It is not widely recognized that zero also has an essential ambiguity in its meaning. Normally it operates as a "counter" in a place, like any other digit; thus 403 shows that there are no tens to be counted, just four hundreds and three units. But when we write 100, it is not immediately clear whether we are referring to $99 + 1$, or to a single larger unit of one-hundred, analogous to a dozen or score (meaning twenty, as realized in the French "quatre-vingts" for eighty). This ambiguity is not an unfortunate accident; it is essential for the conceptual coherence of the place-value system. Thus a thousand, as a larger unit which is capable of being multiplied by digits from 1 to 9, needs three "filler" digits for its expression. (Zeroes immediately to the right of the decimal point have no such ambiguity, since they define sub-units in a context of measurement rather than of aggregated counting).

In standard practice it is sufficiently clear from the context whether zero is a counter or a filler. So much so, indeed, that the distinction is hardly noticed. The fossils joke depends on the attendant knowing only the standard arithmetical rules, and being unaware of the ambiguity of the zero. What he "should" have done was to spot the ambiguity in the symbolic language, to make a judgement on the meaning appropriate to the context and then to decide whether to over-ride ordinary arithmetic. His failure to do so did not result in a reactor core melt-down, but only a joke. We see here how the skills of recognizing ambiguity and over-riding standard rules interact in practice. A closer look at the expression for fifty million shows how the ambiguity of zero cannot always be simply resolved. In 50,000,000 there are seven zeroes after the digit five. How many of them are counters, and how many are fillers? The correct answer depends on how the numerical expression is to be qualified, both in terms of its characteristic uncertainties and in terms of its intended use. We will discuss this at greater length later.

3.8. ROUNDING-OFF: THE π-DILEMMA

Complementary to the "fossils-joke" we have the "π-dilemma". This concerns digits to the right of the decimal point, where they express sub-multiple units. It is significant that "decimal fractions" are an extremely recent invention, appearing for the first time in a little book by the Flemish scientist Simon Stevin in 1585. There he extended the place-value system to give a recursive sequence of sub-multiple units, so that ordinary arithmetic could be extended to that case. Stevin's sub-multiples are normally related to continuous measurements rather than to discrete counts. For any quantity as expressed, only a certain number of places (related to the extent of inexactness) are meaningful; and the last place represents an estimate which is necessarily inexact. In arithmetical operations with decimal fractions, there is a constant danger of producing nonsense. For if we multiply two numbers given to thousandths, the product extends to millionths. Worse yet, the quotient of two simple integers can be an expression that goes on endlessly, as $1/3 = .333....$ Thus we see that the language of this slightly enriched arithmetic (integers plus decimal fractions) would, under very simple operations, yield meaningless expressions. This is an even more severe problem than the ambiguity of zero discussed above. Therefore, the language of arithmetic as extended to this case needs to be controlled by special rules that constrain the ordinary operations. Normally this is accomplished by the significant digits and rounding-off conventions, which govern the rounding-off, or discarding of extra places. These prevent hyper- or pseudo-precision in the outcome of calculations. There is an artefactual arithmetic for multiplication and division, where the number with the least digits sets the standard: all others carry at most one more digit than it, and the answer is then rounded-off to that same number of digits. By such an artefactual arithmetic, the production of pseudo-precise numbers by calculation is prevented.

Such simple rules serve well for the majority of cases; but if one enquires more closely into the behaviour of the uncertainties as they are propagated through the calculations, a need for higher-level rules, applied by judgements, becomes apparent. For we find cases where it is not obvious whether a particular decimal digit is truly "significant" and therefore to be retained. Here we have an example of a vague boundary between those symbols which are meaningful and those which are not.

We can illustrate this point by the "π-dilemma". Suppose we are calculating the semi-circumference of a circle. How many decimal places of π should we use? It is normally assumed that one should round-off the more precisely known number until it has one more digit than the other, and *then* operate. The answer is rounded to as many digits as the less precise number; and all the uncertainty in the answer is due to its "tolerance". But a consideration of two cases shows that this may be an ineffective procedure. First, the radius is assumed to be 1.2 units; as it is not 1.1 or 1.3, its tolerance is $\pm 8\%$. Then, if we round off $\pi/2$ to 1.57; its tolerance is $\pm .6\%$; and so it is appropriate. But if the radius is 9.2 units, the tolerance is little more than $\pm .1\%$, and so $\pi/2$ should be rounded to 1.571, *two* extra digits, lest the tolerance of its rounded-off expression be of the same size as that of the measured radius. Hence the appropriate length of the string of decimal places for $\pi/2$ depends critically on the context: the values of the other elements of the calculation. One might imagine a supplementary rule for significant digits in which such extreme values were distinguished. But then there would need to be a division point between them, as at 5, perhaps; and doubtless more supplementary rules for the more refined case. Since the proportional tolerances of the numbers will never be exactly represented in the product, being too large or too small, there can be no perfect rule. Thus there is no simple rule for resolving the ambiguity in the "significance" of any given place in the decimal expansion of π. The dilemma can be resolved in practice by rules of thumb which balance accuracy of representation against convenience.

3.9. CRAFT SKILLS AND "MONSTERS"

The "π-dilemma" is an example of what happens when a mathematical language is extended to be applicable to new sets of objects and uses. There appear instances of what Lakatos called "monsters" (Lakatos, 1976). The infinitely recurring decimals are a simple example of this; the "π-dilemma" is more subtle. In this case, it is the use in the context of multiplication, of implicit conventions developed around linear measurement. In multiplication, it is proportional errors which are relevant; and these are not managed by the linear system. This dilemma is a very elementary form of one with which many scientists are familiar: which statistical tools to apply to information in logarithmic form that is derived from linear data? Thus a least-square regression on data plotted on a semi-log graph requires a sophisticated interpretation to be valid. Our point is that the choice of an appropriate mathematical language depends on context and use; these examples show that when these are mixed there will be incongruities within the language as applied.

It is noteworthy that the "π-dilemma" is not a joke. This shows us that a rough and ready craft practice of managing decimal expressions can and does exist, with no obvious signs of problems or harmful effects. Among experts in applied statistics, certain rules are indeed known, for rounding-off the results of samples. These are framed for a combination of plausibility and convenience; and their interpretation depends on operators' judgements of the context (see, for example British Standards Institution, 1975). The lack of awareness among non-specialists of the problems of the inexactness of arithmetical expressions, will not usually manifest itself in ordinary practice. But there are areas where that practice is inadequate to the task, and unselfconsciously applied craft skills do not protect against statements which are scientifically meaningless or misleading for policy. There we need an improved notational scheme explicitly designed to guide and enhance the skills of management of uncertainty.

What we have said about mathematical language may seem to run counter to all the commonplaces about mathematics that have been articulated by philosophers, purveyed by teachers and absorbed by students for a very long time. From the time of Plato onwards, mathematical language has been prized because of its supposedly complete freedom from ambiguity and vagueness. Now we are arguing that these attributes are ever-present, and indeed essential for the power and creativity of mathematics. Relevant to our present work, we are showing that the management of uncertainty depends essentially on an appreciation of these properties of mathematical language. Only then can we develop the skills necessary for "knowing-how" to manage uncertainty, and so accomplish the assurance of quality of quantitative information which is so urgent in the areas of policy-related research.

CHAPTER 4

CRAFT SKILLS WITH NUMBERS

Our discussions of uncertainty have shown how existing numerical systems, and the procedures for their use may be inadequate for the new functions of its management. The systems themselves cannot provide explicit rules whereby all instances and exceptional cases are encompassed in advance. The interpretation of new situations in terms of existing rules, and the decisions on when to over-ride those rules, must be made on the basis of judgements. If these are to be effective, they must derive from some form of knowing. This knowing, unlike that of the philosophical ideal of logic and mathematics, will be partly tacit, based on a broad experience, and realized in practice through skills. Such a way of "knowing-how" is not opposed to, but rather is complementary to the explicit "knowing that". Thus skills are necessary in any application of a mathematical system to the real world. (For a full discussion of skills in scientific research, see Ravetz, 1971, 75–108.)

4.1. SKILLS IN STATISTICS

The need for craft skills is well illustrated by the example of statistics. Statistics is not a branch of pure mathematics, but rather a set of conceptual tools whose functions include the representation of the properties of sets of data, and the assessment of their quality. There has been a tendency to present statistics as a mathematical technique into which some unproblematic data are introduced, or (more recently) as a computer package requiring little understanding for operation. But as we mentioned already, problems of data-quality, and of comprehending the structure of the underlying process, may be more critical to the exercise than the numerical calculations performed by the program.

Further, much of statistics incorporates value considerations into its mathematical techniques. At its most familiar, this is revealed in the pre-assigned confidence level for standard tests of significance. The choice of level depends on the assumed relative costs of the two possible sorts of error (which for simplicity may be called false-positive and false-negative). As an extreme example, quality-testing should be less rigorous on crates of apples than on crates of land-mines; one rotten apple will spoil only its own barrel, while a hyper-sensitive land-mine can take the whole warehouse with it. Since uncertainty can never be completely abolished, and also since there are always constraints on resources for testing as well as for production, the statistician (or in general the scientific researcher) must make a judgement to balance the

49

various costs and benefits. This cannot be done in a purely quantitative way, as value-commitments are involved. There must be a skill for such sensitive work, more refined than the simple technical operations.

Value-commitments enter the testing procedures in ways that are far more intimate than is commonly realized. For example, in tests for cancer in animals, only malignant tumours are counted. But as A. S. Whittemore observes, "the line between benign and malignant tumours is a fine one, and the extent to which a pathologist's values affect his decision is unclear". In this context "unclear" means "unknowable", unless this sort of "personal equation" were to be assayed by expensive tests. Hence the basic data on carcinogenicity, on which so many regulations depend, incorporates this unquantifiable, subjective element.

Another influence of values is in the balance of attention given to possible false negatives and false positives. Whittemore provides two examples of bias in opposite directions:

In every animal bioassay there is the probability that a noncarcinogen will, by chance fluctuation, test positive. We can lower this false positive probability by reducing the bioassay's power to detect a true carcinogen – a value-laden tradeoff. Page (1978) has argued that the low statistical power of many toxicological studies effectively protects chemicals more than people. Conversely, the opening paragraph of each NTP–NCI bioassay report (National Cancer Institute, 1981) ignores the possibility of false positives. Negative results, in which the test animals do not have a significantly greater incidence of cancer than control animals, do not necessarily mean the test chemical is not a carcinogen because the experiments are conducted under a limited set of circumstances. Positive results demonstrate that *the test chemical is carcinogenic to animals* under the conditions of the test and indicate a potential risk to man (italics added).

The asymmetry between false negatives and false positives in this paragraph reflects a value judgement. (Whittemore, 1983, 27.)

Further, in the use of statistics in policy related work, the data are frequently so sparse and so weak that the ordinary techniques do not produce conclusions of the necessary strength. Even in classical statistics (as an aid to statecraft) the problems of aggregation can be resolved only by skills and informed judgements. Thus in international statistics, data of a region of the word may be completely dominated by that for a single country: thus in South Asia, India dominates strongly or even overwhelmingly over all its smaller or poorer neighbours. One can well imagine the frustrations of a statistician who has really good data from a constituent that makes a small relative contribution to the aggregated total; how aggravating to discard it because of its being swamped by the size and even the uncertainties of data from a less cooperative but dominant constituent. In the management of such issues are deployed the craft skills, involving both calculation and representation, of the statistician. (See, for example, World Health Organization, 1984a, 5.)

Unfortunately, such skills cannot be taken for granted; and worse, the recognition of their necessity is inhibited by the prevailing metaphysical assumptions about the certainty of information presented in numerical form, inhibits the recognition of their necessity. On occasion, this lack of awareness produces blunders among otherwise competent experts, which seem like the

TABLE I
Table of arisings (Hazardous Waste Inspectorate, 1985)

District	Arisings (tonnes)
Cheshire*	259,000
Cleveland	119,178
Cornwall**	67
GLC***	269,000
Wiltshire	2,770

* Estimated figure as WDA's records are not in this format.
** Estimated figure as WDA's could not supply figures.
*** Estimated figure as WDA's refused to supply figures. (Some WDA's refused on political grounds to cooperate in any way with the HWI.)

fossils joke brought to life. For an illustration of this phenomenon we discuss a set of published tables, part of the first report of the Hazardous Waste Inspectorate (HWI) of the United Kingdom. The HWI requested figures for "arisings" (production plus imports minus exports) of hazardous wastes from each of the Waste Disposal Authorities (WDA) corresponding to local government districts. A few entries from the table of arisings are displayed in Table I (Hazardous Waste Inspectorate, 1985).

All the numbers in the table were summed; the total for England being given as 1,590,014 tonnes. What does this sum mean? More than a third of the total tonnage (coming from the GLC – Greater London Council – and other major industrial areas) is qualified as *** information, and (with one exception) is given with "000" in the last place. When we see such uniform strings of zero digits at the end of a numerical expression, we assume that these are "filler" digits, rather than "counter" digits expressing the result of physical or mathematical operations. When there is a string of three of them, they represent a unit of aggregation in thousands; in those entries, numbers less than 1,000 are strictly meaningless. Yet other entries in the tables, ranging from Cornwall to Cleveland (with its six-digits precision), are all added to these, yielding the final sum, itself expressed a precise to seven digits. Clearly, in the arithmetic itself, the whole Cornwall entry is swamped by the inexactness of the others. What makes the calculation more remarkable is the clear and perceptive introductory discussion in the Report, concerning the extreme unreliability of *all* statistics on arisings of hazardous wastes. One may ask why the arisings were not recorded to the nearest kilo, since tonne is really not much more meaningful. Such questions can be very useful in the elicitation of uncertainties, as we shall see later.

This is not to say that such a table of arisings is useless or misleading in itself; much valuable information on the general size and location of the problem, as well as on the state of statistics, can be obtained from it. But what conceivable policy decision could depend on the pseudo-precise calculated sum 1,590,014? A more genuinely accurate summary of the table would highlight the general size of the various categories of the information; and their associated uncertainties. Thus we might say:

Totals of estimated arisings: 10^6 tonnes.
Totals of reported arisings : 6×10^5 tonnes.

Standard notations go some way to help avoid such confusions as we have seen in the table. Thus, for example, we may write,

Cheshire* 259K
GLC*** 269K

If such aggregate units were used, the precision of the Cleveland entry, given to one part in 10^6, would call out for attention and critical scrutiny. In general, if some entries are given to the nearest thousand, then it is clear that the smaller place-values of the others are swamped in any sum. With such principles in mind we might look again at those entries are expressed to the nearest K-ton. Is there any justification for a precision of within 1 %, when records were either in a different format, or not even supplied? It would seem that a more flexible system of notation for estimates would be useful here. Thus both the Cheshire and the GLC entries would be well expressed as $2^{1/2}$: E5-tons, where the En is the standard scientific notation for 10^n. Although $2^{1/2}$ is not standard, it certainly conveys the rough estimates better than any other set of digits. We will show how such notations can be incorporated into the NUSAP notational scheme.

The damage done by such excessive precision is not always restricted to the reputations of those responsible for it. Inputs for policy decisions are quite commonly expressed in mathematical language. This is done even when they express only preferences and values, for which an appropriate topology would be very coarse indeed. As a consequence, decisions which *seem* to be derived from quantitative facts and scientific method, may actually be determined by chance, blunder, caprice, or covert politics. Paradoxically, if decision-makers overturn the conclusions of such studies in favour of overtly political considerations, they may actually be increasing the "objectivity" of the outcome.

4.2. SKILLS IN COST-BENEFIT ANALYSIS

An illustration of this point is the Cost-Benefit Analysis (CBA) for the siting of the proposed Third London Airport in the 1960's. Here a long and expensive study was conducted; the key indicator was the estimated "total resources cost" over a twenty-year period for each of the proposed sites. The two leading options had cost estimates differing by about £100,000,000 Sterling, a large sum of money in those days. But this was only a difference of less of 3 % between the two calculated costs, of which the larger was £4,250,000,000. Also, it was possible for critics to find somewhat arbitrary elements in the calculations, which accounted for much of the difference. One such element was a supposed five minutes difference in travelling time to Central London, which when costed for many millions of travellers, constituted half of the difference. When the scientific standing of the CBA was so easily discredited by critics, a political campaign was able to reverse the decision (Dasgupta and Pearce, 1972, 201–234).

The choice between the options depended not on the quantities themselves, but on their differences. We may start our analysis by considering how the uncertainties of the terms of a subtraction-sum influence the uncertainties in the result. Because in this case the numbers were constructed out of so many sources and by so many different methods, it is not easy to specify their spread. According to one one school of thought, if the uncertainty is unknowable it should be neglected. But this is erecting ignorance into a principle of action. Since one straightforward source of uncertainty already amounted to half the difference, we may say that the spread of the difference runs at least from zero to twice the stated $£10^8$.

The debate over the results of that cost-benefit analysis were very instructive, for it showed that the skills of management of uncertainty that are taken for granted in any matured quantitative science are simply unheard of in many others. There is really an enormous gap between those who are familiar with these problems and those who are not. There may well be some who still canot see what is wrong with the stated $£100,000,000$ difference (it does after all represent a lot of money); and there may also be many who do not notice the incongruity of those long strings of zeroes.

What would be a sensible expression for this "total resources cost"? Put otherwise, how many zeroes should we lop off, as being fillers and not counters? Anything less than $£10^8$ is not significant here. Accepting the precision of the stated estimates, we can express them as $41\frac{1}{2} \times £10^8$ and $42\frac{1}{2} \times £10^8$. Then if we express the difference-sum formally, as

$$
\begin{array}{r}
42\frac{1}{2} \times £10^8 \\
-41\frac{1}{2} \times £10^8 \\
\hline
1 \ \times £10^8 ,
\end{array}
$$

it is easily seen to belong to the same general class as the "fossils joke".

Since then CBA has gained much greater sophistication; but it is still vulnerable to pseudo-precision. A more recent example, is provided by an application of the technique to Health Economics, for a major health service rationalization. Table II shows the Costs and Benefits of the different options considered. There is an explanatory paragraph:

Option 6 (Do Nothing) has been rejected because of its unacceptable low score. Option 1 has been selected as the preferred solution because it is the cheapest of the remaining options and has a benefit score close to the highest. Its service superiority is found to be sensitive only to changes in weighting of criteria which would be contrary to current strategy. (Akehurst, 1986, 5.)

TABLE II
CRA of options for a health review (Akehurst, 1986, 5)

	Annual equivalent cost of capital and revenue together (£ thousand)	Benefit score
Option 1	38,425	702
Option 2	38,534	707
Option 3	38,600	688
Option 4	38,815	469
Option 5	38,559	675
Option 6	38,185	322

There is also some discussion of the uncertainties in the estimations, as:

It should be noted that the differences in revenue costs are small in general, and error in estimation would account for a good part of the differences. In addition, the differences are very small in comparison to the total savings in revenue that all options would be expected to achieve in their steady state. (Akehurst, 1986, 47–48.)

There is no sensitivity analysis on the effects of the assumed discount rate (set at 5%), which would be very strong in the present case. Worse, the Benefit Score, derived from individual scores on a scale from 1 to 10 which were multiplied by weights representing policy priorities, is hardly a quantity significant to three digits. (There are also noticeable arithmetical blunders in the table of aggregated non-financial benefits) (46).

What significance can there be in the differences between the three first options, less than 1% on the financial side and some 3% on the non-financial side? Were there to be error bars for inexactness, these quantitative differences would surely be swamped. Do the practitioners of CBA believe that real scientists would consider such statistically insignificant differences to be genuine? Such examples show how some decision-makers must still be mesmerized by "magic numbers", for these exercises to be taken seriously.

4.3. SKILLS IN SCIENCE

Examples as these lead us to appreciate a new sort of pseudo-science based on the magic of numbers and the neglect or ignorance of craft skills. This is defined by its methodology, in the unavoidably uncritical and unskilled use of mathematical language. Such a modern pseudo-science may be defined as one where the uncertainty of its inputs must be suppressed, lest they render its outputs totally indeterminate. A convenient name for it would be "GIGO-Science", as an extension for the familiar American acronym for the misuse of computers: "Garbage In, Garbage Out". How much of our present social, economic, military and technological policies make essential use of GIGO-Sciences is one of the more important questions of our age. Fortunately, the GIGO-Science component in decisions seems to be used more frequently for rhetoric than for substance.

The reason why such pernicious practices can flourish is the widespread ignorance of the nature of mathematical language. This results from the perennial domination of the metaphysics of mathematics by a paradigm which imagines it only as an ideal conceptual structure, suppressing its complementary aspect as an abstracted practical tool. That one-sided paradigm, in whose terms uncertainty cannot be managed, is now in crisis. We are showing how with appropriate notations, quantitative expressions can embody the open-endedness that is necessary for the development of their fruitful contradictions. Their attempted suppression is as unnecessary as it is harmful. With this new kind of mathematical thinking, we can help in the development of techniques and craft skills for quality-assurance of information expressed in mathematical language.

Our task on the philosophical side is to overcome the prejudice, deeply ingrained in our intellectual culture, which makes "knowing-how" an inferior sort of practical knack, unlike the genuine "knowing-that" of erudition and scholarship. The societal context of such a distinction reflects the division of hand-work and brain-work which has existed for millenia. Here we are showing that this divided conception of knowledge is now becoming damaging to all of society, independently of its political, economic and social structures. One way to break down the distinction is to show that even mathematics itself does not maintain it in practice. For in education in mathematics at all levels, a large part of the work is the development and fostering of skills. This indeed explains and justifies the repetitive, frequently boring exercises which students must do. The knowledge gained thereby, almost all of the "knowing-how" sort, is recognized by teachers (if not by philosophers of knowledge) as necessary for understanding in even the most abstract of academic fields. As we have seen, the experience of work in metamathematics shows that the manipulation of symbols in a proof is not an automatic path to understanding. For understanding a proof, it must be re-created by the learner; and there must be informal criteria of choice among possible symbolic moves. Otherwise the process is anarchic and a solution is arrived at only by chance. Each learner must use a problem-solving strategy based on judgements of likely paths forward to a solution. This is achieved by experience and precept, trial and error, complementary to the more formal presentations given in lectures. All this happens even in abstract mathematics, where the empirical content, and relation to general human experience, is minimal. The outcome of such learning processes, incorporating both these complementary aspects of education, is a subject truly mastered. Its resulting knowledge is then available for further development or application. Otherwise, the student who has merely memorized formula derivations has no idea how to apply them outside the special conditions of the examination room.

The craft skills of research incorporate those of the student, and in addition involve those of independent work. Some of these skills are of a "housekeeping" sort, but are no less important or demanding for that. Thus, for example, clean work, appropriate to its circumstances, is essential to prevent false results in some experiments, contamination of samples, and accidents. Similarly, meticulous record-keeping is necessary so that precious data is not lost. It may seem that these are trivial aspects of research work, not worth much reflection or care; but in fact they require great skill, since perfection is impossible either in cleanliness or recording. Excessively high standards can be harmful, in their diversion of energy and attention; and being essentially pointless they eventually become a matter of bureaucratic games or personal obsession, or simply collapse. This craft knowledge, without which high-quality research cannot be done, is overwhelmingly of the "knowing-how" sort.

These elementary skills do not exhaust the list; if the management of resources of materials and manpower is involved, administrative skills are also required. Modern conditions of scientific work demand new skills as well, such as the humane treatment of animals, and the prevention of accidents when dangerous materials (such as radioactive substances or genetically engineered organisms) are used, both within administrative frameworks set by law; to say nothing of the skills required for writing grant proposals, an activity which seems to occupy an ever increasing proportion of the research-ers' time. There are also the new skills of communication to colleagues or peers in other fields. This is particularly delicate task when the communication is to lay publics, through the media. Finally there are the skills which distinguish a genuine investigator from a research-worker. These require the initial selec-tion and framing of problems, and the guiding of their evolution through the course of the research process. At the community level in a field, there are the social skills of setting the criteria of adequacy and of value on which all the judgements of craftsmanship and quality are based.

4.4. DEGENERATION OF SKILLS

European civilization is distinguished among all others by the mathematical ideal of knowledge, which has always permeated our physical science and technology. This has been, in many respects, an overwhelming success. Policy-related research reflects the phenomena that display the negative side of this achievement. It provides us with a strong base in experience for criticizing this traditional one-sided conception of knowledge, where 'know-ing-how' has always been displaced from the focus of reflection on knowledge and practice. As a result of this bias, the transmission of skills, without which there can be no enduring quality in science (or in technology), is neglected in philosophic theories and even in the self-awareness of science. The way in which skills are learned in science is hardly different from the procedure in traditional crafts, where apprentices learned from the masters by imitation and precept. Such a transmission of skills is done informally, usually even unselfconsciously, and is therefore very vulnerable to degeneration. For when a scientist has research students who are either excessive in numbers, or defective in motivation, this essential core of research training is omitted, without anyone being aware of the loss. The next generation of scientists, not knowing what they are missing, will provide an even more defective training, and so on. In not many generations, the activity of research is very different from what it had once been.

The process of degeneration of skills is accelerated when new tools for the automatic creation of data are available and amenable to uncritical use. These (as for example, computer packages for statistical treatment of data) are generally both labour-saving *and* opaque, tending to replace skills instead of increasing and complementing them. Why such tools work as they do, and how one can spot the pitfalls of inappropriateness, or even of simple error, are issues which can easily be ignored by researchers concerned only with getting out some plausible numbers. When the researchers themselves feel unsure of their mastery of the numerical or symbolic skills, then these automatic methods provide a means of escape from an incomprehensible and threatening situation. All the necessary skill in mathematics can now be transferred to the machine, providing security at least in the short run. Then the scientists and their research students are relieved of any further responsibility for mathematical competence of any sort. This situation is especially dangerous if there is an attractive multiple-colour 3D graphic package for visual displays; who will dare to make the distinction between the quality of the representation and the quality of the contents?

As automatic systems become ever more powerful, the tendencies to de-skilling of humans can extend even to depriving them of their elementary competence, and their ability to recognize the state of affairs that is affecting their practice. For, as expert systems become more widely used, humans would be progressively displaced from intellectual work as well as manual; and no-one could really know what the program is doing except for the software engineers who wrote it, and who have since then moved on to another job. Such a system is vulnerable in many ways; most clearly, if it "crashes" then the relevant knowledge, social structures, and institutional activities go with it. Much more subtly, the expert systems may come to be used as the standard for competent practice, so that any human deviating from them by their personal judgement would be at risk (Spencer, 1987). We have already seen in Chapter 2 that the tendency to uncertainty-avoidance by the reification of policy-numbers is an entirely natural one in the context of bureaucracy. From this tendency to aggravated de-skilling by computer, could come a rigidity and vulnerability no less dangerous and destructive in the long run in spite of its apparent sophistication. A true marriage of intellectual and practical craft skills with automatic conceptual tools is thus one of the more urgent tasks for maintaining and improving the health, performance and quality of our technological enterprise.

Another contributory factor in the tendency to the dilution of quality in science is the increased specialization of research. In Kuhn's, terms "normal science" consists of "the strenuous and devoted effort to force nature into the conceptual boxes provided by one's professional education" (Kuhn, 1962).

This education, which Kuhn compares in its narrowness to orthodox theology, afflicts mathematics, natural and social sciences alike. Moreover, the loss of a broad cultural basis attendant on the rise of mass education, isolates the technical core of science from its context, and also from its humanistic aspects. Then critical reflection on practice, on philosophical aspects of scientific research, and on its place in society, are impaired by such a narrow training and indoctrination. Those scientists who have such broader concerns are at risk of being disapproved by their colleague. Such diversions from the "real" work of research and grant-getting are costly to them not only in lost time but worse, in a damaged reputation. Is this an inevitable by-product of the industrialization of science and of the extension of access to higher education?

Thus another great challenge of our time is to recover the sense of quality in science, without recourse to an elitism that in any event is no longer possible (Ziman, 1960). The age has long since gone when a class distinction was made between (amateur) "gentlemen" and (paid) "players" in English cricket, who were provided with separate changing rooms, and when the "captainship" of the team representing England was reserved for a "gentlemen". The attempted translation of the squirearchy of the village to a mass-media spectator sport could not be sustained. By analogy, the "little science" of the lone researchers of modest but independent means, with their low-technology apparatus and small clubs of colleagues, is now negligible. Can the commitment and morale of those earlier heroic days of science, so necessary for the maintainance of quality assurance, be sustained when its social basis has been transformed out of all recognition? (For a full discussion of quality-control in science, see Ravetz, 1971, 273–288).

Thus science is an ongoing process, and not tables of enshrined truths. Philosophers who have concentrated on the intellectual product of science and ignored the social process, have too easily imagined the knowledge as timeless. The history of science provides us with illuminating examples of what happens to science when it is deprived of the stimulus of the new research, and its reinforment of quality. The scientific literature of Rome, aside from Pliny's *Natural History* and Lucretius' *On the Nature of Things*, was a product of a culture where research was never prized. There "knowing-that" was completely dominant over "knowing-how", and the long term effects were destructive of skills as well as knowledge. It left a record of nearly unrelieved banality, plagiarism and degeneration (Stahl, 1962). By contrast, in the Eastern Roman Empire the original Hellenistic traditions of research, as in astronomy, mathematics and medicine, were maintained, and scientific excellence flourished for centuries later.

4.5. POLICY-RELATED RESEARCH AND SKILLS

A focus for a response to both these challenges, the enhancement of the skills of research and the protection of its quality, is provided by policy-related research. By definition, policy-related research is research in aid of policy. A research exercise conducted in ignorance of, or indifference to, its stated policy context will be useless at best. This is not to say that basic or even pure research is necessarily to be dismissed as irrelevant; but that when the results of such research are brought into the policy arena, they need reinterpreting in terms of this additional dimension. This context will not be the same as the erudite culture of the elite academic education of previous ages; but it involves its own sort of broad, reflective outlook.

Similarly, policy-related research cannot be conducted effectively by a return to the narrow style of Kuhn's "puzzle-solving within a paradigm" of "normal science". As we have seen, the uncertainties are so severe, and so permeate the whole work, that only highly skilled and reflective judgements are appropriate for their proper control. For here, the maintainance of quality (as "goodness" of the product of research) depends crucially on the management of its uncertainties. But these cannot be grasped in isolation of the intended policy context of the work. For the anticipated uses of the results condition the judgements of what sorts and what degrees of uncertainty are acceptable. In this way, the quality of the research result considered as an input (analogous to a tool or component) is assessed by its performance of its function, and is closely related to that "quality" which characterizes the craftsmanship of the research itself.

Policy-related research requires, and therefore stimulates, the rediscovery of craft skills as a genuine constituent of scientific work, and indeed as a genuine form of knowledge (Polanyi, 1958, Pirsig, 1974). These will include the traditional skills of research technique itself (for these may also be easily lost), those of quality assessment, and now the new skills for the communication of uncertainty. In policy-related research the tasks of communication are more demanding than in traditional scientific activity, because an inexpert user of the research results is frequently part of the problem as formulated. In this context, "popularization" is not an afterthought to the research, or a way to secure research funds, or a propaganda exercise for Science. What the researcher provides to users will not be for their amusement or edification, but an important component or tool for a process, with mixed political and intellectual aspects. Hence the new skills of communication of scientific results are like a translation exercise.

This translation exercise is less like that between English and French, and more like that involving Japanese: the incongruous cultural contexts render word-for-word translation futile, for too many concepts lack significant common elements in their families of meanings. In the present case, this is partly on the technical aspects, but it also affects uncertainty. For the research process is focally concerned with cognitive uncertainty (even though this is conditioned by values, as in the prior setting of confidence-limits in statistical tests). This permits the qualification of conclusions, even when relevant to policy, by the use of expressions such as "provided that", "other things being equal", "yes and no with reservations", and so on. This is not appropriate for decisions; such qualifications arouse reactions like that of the American congressman, who called for a one-armed scientist, who could not always say, "on the one hand..., on the other hand...".

Decision-makers involved in policy or politics are no strangers to uncertainty. But their uncertainty concerns an intuitively sensed context of unpredictability and unreliability, where survival is the objective and the task is not a science but an art. For them, the quality of information inputs tend to be assessed pragmatically: will they help me win, or deflect blame if I lose? Thus the criteria of quality relevant here are complex in themselves and very different from those at the research end. (See for example W. C. Clark and G. Majone', "Critical Criteria", Table III.) Further, the sorts of scientific uncertainty which are relevant in this policy context may be very difficult to grasp. The very small probabilities cited in the risk-related literature are alien to human experience. W. D. Ruckelshaus provides the following example:

Tell somebody that their risk of cancer from a 70-year exposure to a carcinogen at ambient levels ranges between 10-5 and 10-7, and they are likely to come back at you with: Yeah, but will I get cancer if I drink the water? (Ruckelshaus, 1984.)

It is not merely decision-makers and lay persons in general who are baffled by expressions of uncertainty associated with policy-related research. Even scientists themselves may find it difficult to comprehend the uncertainties, and therefore to evaluate the quality, of results in specialities other than their own. Multidisciplinary teams then generally become a collection of specialists playing safe by abstaining from criticism of others' research results. A questioner can all too easily be driven off and humiliated by an aggressive defence; and so in the absence of a common understanding on the issues of uncertainty and quality, it is futile for an expert to attempt to stray onto another's turf. Since policy-related research involves complex systems which have been approached from a plurality of disciplinary perspectives, this systematic weakness of multidisciplinary projects must be resolved, if effectiveness is to be achieved and progress to be made. Thus we see again that the improvement of communication of uncertainty is important for research science no less than for an inexpert lay public.

4.6. NEW SKILLS FOR POLICY-RELATED RESEARCH

Policy-related research thus requires the development of new skills, for the effective performance of its variety of new functions of quantitative information. Among these functions and associated skills, we can discuss "indicators", "policy-numbers", "forensic science", "communication", "data-base quality", "elicitation of uncertainty", "information evaluation", and finally "uncertainty management in debate".

Much of policy-related research is devoted to the tasks of managing a complex environment, natural and social. There is no way that this can be described in full detail; what is needed for policy purposes is a set of "indica-

TABLE III

"Critical criteria" (Clark and Majone, 1985)

Critical role	Critical mode		
	Input	Output	Process
Scientist	Resource and time constraints; available theory; institutional support; assumptions; quality of available data: state of the art	Validation; sensitivity analyses; technical sophistication; degree of acceptance of conclusions; impact on policy debate; imitation; professional recognition	Choice of methodology (e.g., estimation procedures); communication; implementation; promotion; degree of formalization of analytic activities within the organization
Peer group	Quality of data; model and/or theory used; adequacy of tools; problem formulation; input variables well chosen? Measure of success specified in advance?	Purpose of the study; conclusions supported by evidence? Does model offend common sense? Robustness of conclusions; adequate coverage of issues	Standards of scientific and professional practice; documentation; review of validation techniques; style; interdisciplinarity
Program manager or sponser	Cost; institutional support within user organization; quality of analytic team; type of financing (e.g., grant versus contract)	Rate of use; type of use (general education, program evaluation, decision making, etc.); contribution to methodology and state of the art; prestige; can results be generalized, applied elsewhere?	Dissemination; collaboration with users; has study been reviewed?
Policy maker	Quality of analysts; cost of study; technical tools used (hardware and software); does problem formulation make sense?	Is output familiar and intelligible? Did study generate new ideas? Are policy indications conclusive? Are they constant with accepted ethical standards?	Ease of use; documentation; are analysts helping with implementation? Did they interact with agency personnel? With interest groups?
Public interest groups	Competence and intellectual integrity of analysts; are value systems compatible? Problem formulation acceptable? Normative implications of technical choices (e.g., choices of data)	Nature of conclusions; equity; analysis used as rationalization or to postpone decisions? All viewpoints taken into consideration? Value issues	Participation; communication of data and other information; adherence to strict rules of procedure

tors", whereby crucial aspects of the total situation are identified amd their state evaluated. The quantities appearing in that role may seem to be like those numbers emanating from traditional laboratory research; but the resemblance is only apparent. For such indicators are highly artefactual (which is not the same as being fictitious or totally arbitrary); they relate back to empirical findings through a complex structure of special policy-driven definitions, conventions for statistical accounting, and selected samples for data-collection or research. The general public gets a hint of their character, and of their inherent uncertainties, only when some controversy forces this out into the open. The only indicator which regularly exhibits its uncertainties as part of its quantitative statement, is that of public-opinion polls! This scrupulousness may be related to their vulnerability to the dissatisfaction of their clients, for whose custom the different polling organizations compete. In addition to the standard skills of statistics, indicators require an enhanced awareness of the significance of the background policy commitment, especially as it shapes the categories of the data. Also, there is a skilled design exercise in setting the degree of quantification, as this is reflected in the coarseness of the scaling of the indicator.

Analogous to indicators, but even further removed from "science", are the numbers used for regulation purposes, typically for defining a limit of some sort. A limit may popularly be thought as defining "safety"; or it may be a threshold for an alert or for a prohibition, or perhaps even for a prosecution. Such "policy-numbers" have some factual basis to be sure; but the assignment of their actual value (out of a range of credible values) will depend as much on administration, negotiation and pressure-group politics, as on firm scientific results. Hence these policy-numbers reflect fiat as much as fact; and as their origins and functions are different from those of numbers as traditionally used, so their form of expression could well be reconsidered. The production of these policy-numbers requires some skills which are very new indeed for the scientists. Researchers are not strangers to scientific politics; but it has always been an article of faith that the consensus on scientific results must be totally independent of extraneous considerations. In dealing with policy-numbers scientists must enter into a process which by their inherited standards may be seen as essentially corrupt. Radically new attitudes and skills, cognitive as political, are demanded. Again, the communication of policy-numbers presents unprecedented problems. Here there should be a balance between precision of statement (for both administrative convenience and expert credibility) and clarity about uncertainties (reflecting the political realities of the decision process). The development of skills for this new task may safely be said to be in a rudimentary state.

Analogous in many ways are the challenges of "forensic science", as when an expert appears in a tribunal of some sort, be it a court of law or a public enquiry. Many researchers have been shocked to discover that their scientific credentials provide them with no credibility at all, at least in confrontation with opposing advocates. The inquisitor has no collegial loyalty whatever to the scientist, and no compunction at destroying the expert's professional reputation at the turn of a phrase. The relevant defensive skills here are largely rhetorical; they are based on the realization that the scientist's assertions are testimony, not facts. The skills of management of uncertainty in these contexts becomes crucial, as this is frequently the most vulnerable element. This is accomplished in a totally different way from science; here it involves balancing of probabilities, burdens of proof, and evaluations of evidence (based as much on demeanor of witnesses as on strength of findings); and finally all this accomplished in an explicit, ritualized style, so unlike most of science.

Rhetorical skills are also involved in the "communication" function of numbers, where the audience is a lay public, reached directly in person or through the media. Here too, the change in recent years is reflected in the spectacle of accredited experts being disbelieved and even destroyed, by interviewers or by aggrieved citizens in a NIMBY group. Their research qualifications and academic connections may be disregarded; the common questions now are: "Who's paying you?" or "Why should we believe you?". Experts exposed to such treatment must learn, frequently the hard way, new rhetorical skills. Overcoming an initial distrust requires an abandonment of the traditional patronizing assumption that the scientist has a monopoly not merely of knowledge but also of rationality. Along with this new awareness comes the recognition that non-scientific aspects of a policy issue may be not only inevitable, but also legitimate and fully rational. To achieve such attitudes in spite of the conditioning of years of scientific education, is a task involving commitment and skill.

The scientific inputs to all such processes should derive ultimately from some sort of research. But for any given policy problem, the data inputs will be enormously various. In their sources they will range from reviews of reported experimental results, down to experts' opinions; in provenance from internationally accredited journals to in-house reports; and in relevance from close fit to the problem at hand, to loose analogues. Also, although science as a whole is a long way away from the "Monk Tetzel" syndrome that triggered off Luther's rebellion, still the phenomenon of "Science for Sale" by consultants in policy-related fields is causing concern (Stutz, 1988). The policy-related researcher requires, in these circumstances, a guide to the quality of those items, so that given some skill in assessing the criticality of different inputs, they can identify those most in need of checking and revision. Unfortunately, such information on quality is usually obtainable only through personal contact with the originator of the data; and that can only rarely be accomplished.

The data-bases for policy-related research generally lack any quality assurance; this is in striking contrast with the sophistication and expense of the means of processing them, involving elaborate models requiring great computer power. It is as if we set up a factory for automobiles or space-rockets, assuming that our suppliers could be trusted to deliver components of good quality in the absence of any testing on our part. Hence for the improvement of policy-related research we need new skills for the quality assurance of data inputs. Ideally, every quantitative entry in data bank should be qualified in such a way, that when we know its function for us we can determine its quality. What new skills would be required for this? Certainly some scientists are experienced in quality-assessment; that is at the core of perceptive review-articles. The extension of such skills to all research scientists, and indeed to the whole community of those concerned with such issues, will be enhanced by the provision of appropriate tools and guidance. The main method in the NUSAP system is elicitation, which fosters the skill of reflection on the scientist's own work, so that uncertainties can be characterized, and quality then assessed.

Policy-related research provides opportunities as well as problems; for what is related to policy should be open to the democratic process, involving citizens' participation. The politicization of uncertainty, while causing problems for scientists and administrations, is actually a positive development. For the uncertainty was already there; now it is out in the open, and since it cannot be resolved inside science, it is appropriately dealt with in public forums. The urgent task is the development and diffusion of the skills necessary for this to be done competently. To appreciate the skills involved in this function, we observe that in debates on policy, the numbers have tended to function as "magic numbers" in the sense we discussed previously. They are promulgated in a form that if frees them from all taint of uncertainty. Delivering incontrovertible imperatives, and invested with the charisma of the State bureaucracy that issues them, they seem all-powerful for providing a reassurance of safety. From being measures of acceptable levels of what is "safe", they come to be believed in as *defining* safety, along the lines described by H. S. Brown in Chapter 2. The most important task here is demystification and constructive criticism. This is a task for *all* those concerned with an issue, be they scientists employed in governmental agencies, or in independent institutions, or the technically expert people working in citizens' or intervenor groups.

The skills required start with the basic ones of obtaining relevant information. This is not a trivial task in countries where "Official Secrets" rather than "Freedom of Information" defines the situation. Also, scientific information produced by scientific employees, or even by independent researchers on contract, may become the intellectual property of their employers and *not* traditional "public knowledge". Obtaining such information, as held in private data-banks, may be difficult or expensive. Even when the numbers are obtained, they may reflect fiat or fantasy more than reality; this is most likely in the case of highly aggregated official statistics. Once the numbers are obtained, the next exercise begins, that of unravelling the mixture of commercial, political, administrative and scientific considerations involved in the process of their determination. Both of these phases require skills of investigation, relating to journalism and perhaps even detective work!

Moving now to the context of debate, everyone needs skills of managing their own uncertainties, as well as exploiting those of the others in an appropriate fashion. The etiquette of admission of uncertainties and of contrary evidence depends very strongly on the nature of the forum. In traditional science, as idealized by Popper, scientists should accept and welcome refutations of their theories. Kuhn's image of "normal" science was otherwise; and in the policy context Popperian behaviour would be Quixotic. However, there are real differences between "advocacy" and "adversarial" forums. Recognizing these is an important skill of effective action on policy-related research issues.

We would ourselves be patronizing if we left the impression that "scientists" as a class are skilled in numeracy and "laypersons" not. One reason for the prevalence of "magic numbers" is the lack of awareness among scientists and experts, in many if not most fields, of what numbers mean and can do, to say nothing of symbolic formulae. As B. Turner says,

It is important to spend time on these topics (limits of quantification) because of the peculiar standing of numbers in our civilization. On the one hand we are mesmerised by numbers, even when they are pseudo-numbers, those who deal with them frequently no less than those who are thrown into a panic by them. On the other hand, the general standard of teaching about mathematical issues is so poor that few people understand fully the nature of the properties of the numbers and numbers systems which they are advocating or excoriating. (Gherardi and Turner, 1987, 10.)

The educational task of developing skills for the management of scientific uncertainty is thus a challenge for society in our time, of the same importance as that of universal education in earlier generations.

4.7. DIFFUSING THE SKILLS OF QUALITY ASSURANCE

With the understanding of science as including "knowing-how", we can provide a philosophical setting for the new phenomenon of counter-expertise in policy-related research. For now, official expertise is contested and compromised. In major policy debates, it encounters technically competent criticism from outsiders, criticism which it has not always been always been able to meet. We will show how this counter-expertise is a necessary feature of the new sort of science; and we will then explore its implications for the social relations of science.

For this analysis, we recall one of the many fruitful ambiguities in Kuhn's concept of paradigm, its cognitive and social aspects. For traditional science, there is no clear distinction between them; and so his model of "normal science" uses them interchangeably. In policy-related research, however, the distinction becomes crucial. First, the sciences involved lack a unifying, ruling paradigm that defines puzzle-solving practice to the exclusion of all other considerations. This may at first seem surprising, since the policy problems are frequently raised by the practices of matured sciences or technologies, be they nuclear power, recombinant DNA research, or transport engineering. But (and here is our crucial distinction) the sciences which are required to *solve* the problem are systematically different from those that *created* the problem in the first place. Thus oncology, epidemiology and even radiological protection, are radically different from nuclear physics and heat engineering. We notice that the "sciences of clean-up and survival" are less "matured" cognitively as they deal with the more complex systems found outside the laboratories, and also suffer from an historical deficit of prestige and resources.

Because of this cognitive weakness, combined with more direct relevance to life and survival, these new sciences have a different style both socially and intellectually. First, their results are not all esoteric, requiring "popularization" to an uncomprehending but appreciative lay audience. They directly address the worries of people, as residents, parents and human beings. The problems, and the sciences themselves, are not restricted to the sphere of public knowledge controlled by an intellectual and power elite; they penetrate the domain of the private and the sacred. As a result, these sciences are much affected by a variety of pressures and influences deriving from values and lifestyles. This phenomenon is an occasion for dismay and disgust of traditional scientists whose competence takes them into these dangerous uncharted areas. But these new sciences, however much they may mature technically, can never become "normal" in Kuhn's sense of having a dogmatic consensus that enforces adherence to a closed set of rules for puzzle-solving. For the scientists themselves are people, with homes and families. They may even experience a direct contradiction of their professional and private lives, on their jobs making the problem, and at home wondering how to solve it.

In the social practice of this new sort of science, the expert, interpreting research results to the lay audiences (including decision-makers), is as integral to the activity as the researchers themselves. Furthermore, counter-scientists and counter-experts (the roles may be merged sometimes), representing constituencies outside established institutions are equally necessary. In the new sort of science, they are required for the transmission of skills and for quality assurance of results. For in the case of the new sort of science, who are the peers? In Kuhn's "normal science", they were colleagues on the job, engaged in that "strenous and devoted effort to force Nature into the conceptual boxes provided by professional education". Such peers are still there, as official scientists and experts; and they exercise quality control within the technical paradigm of their expertise. But the problems of the new sort of science are not ones of purely "knowing-that" within stable paradigms; they include "knowing-how", along with broad and complex issues of environment, society and ethics. Hence it is necessary and appropriate for quality assurance in these cases to be enriched by the contribution of other scientists and experts, technically competent but representing interests outside the social paradigm of the official expertise. Since the sciences involved are, as we have said, less matured, and therefore less technically esoteric, adequate technical competence can be attained without the cost of initiation and indoctrination in a Kuhnian cognitive-social paradigm. The counter-expertise thus functions as a first step towards a democratization of science; not a popularization from above, but a diffusion through the broad debate that in other spheres defines a democratic society.

The introduction of counter-expertise into policy-debates does not run smoothly, nor is it even guaranteed to be successful. The tasks of quality assurance are both more challenging and less controlled. Judgements of cognitive uncertainty will inevitably be influenced by the particular agendas of the experts; and functional quality even more directly so. Hence the tasks of achieving any measure of consensus among all sides, even on what may seem narrowly technical questions, can become quite problematic. In the context of policy debates, where decisions may be urgent, and powerful interests may be motivated by totally non-cognitive concerns, the inconclusiveness of scientific debates between experts and counter-experts may come to endanger the whole process of broader participation in policy-making.

The legitimacy of counter-expertise, still far from firmly established, may be threatened by the failure of programmes for its inclusion in the policy process. Official expertise would not necessarily benefit from a return to a more peaceful life. In these new sorts of problems we cannot run for security to highly trained experts like neurosurgeons or airline pilots. Here, they may simply not exist. Hence, if the inclusion of counter-expertise into policy debates fails, the prospect is of a possible loss of a legitimate role for *any* expertise and science, leaving policy decisions once again at the mercy of naked power-politics.

The central issue is quality assurance; and this involves at its core the individual assessments of quality of scientific results. Traditionally, these assessments were made by scientists nearly unselfconsciously, without the benefit of attention or analysis by philosophers. Such benign neglect is no longer appropriate. We are just now at one of those rare junctures in the development of science, where "normal" practice has lost its rigidity, and philosophical reflection can make a direct contribution to the improvement of practice. Here we are calling attention to "knowing-how" and skills not only in the technical practice of research, but also in the higher-level activities of quality assurance. With this sharpened philosophical focus, we can propose a concentration of attention on the skills of the assessment, control and assurance of quality.

All skills are a synthesis of "knowing-how" and "knowing-that". In part they are personal and incommunicable; but they would die out unless they were also transmissible, by example, by precept and by instruction. It is no different with the new skills of quality assurance. In the industrial field, most phases of quality assurance can be conducted as a sophisticated routine. For science, that is less likely; but it is still possible to provide rubrics, guidelines and elicitation procedures, for the expression of uncertainty, for the assessment of quality, and also for the training in both skills.

By this means, we may overcome the contradiction expressed in the paradox of "hard" policy decisions depending on "soft" scientific inputs. We envisage a process of debate and dialogue operating continuously over all phases of a process, where uncertainty, quality and policy are inseparably connected. That the scientific inputs are contested along with the value considerations, should be no occasion for alarm or dismay. This is what it is like, in the new world of policy-related research. Provided that debate is competent and disciplined, the only loss is of our rationalistic illusions. Beyond the inclusion of counter-expertise, we may imagine the diffusion of the requisite scientific skills among a broader population. This seem a bold and dangerous proposal; but then, in their day so were the extension of the franchise to all men and then to women. For the success of any major social policy depends on consensus; and this requires understanding and identification by a broad public. There can be not "blueprint for survival" laid down by experts, official or counter. The only way forward is through the broad extension of competence in the debates on these issues, a competence which will be as much in quality assessment as in the technicalities themselves.

As the problem matures, the elements of a solution emerge. Our high-technology system not only provides opportunities for literate culture for growing sections of the population; it also needs a broadly diffused high level of literacy for its effective operation. Hence politics changes its form; the old oppositions between a rational and sophisticated elite minority and an ignorant, oppressed majority, are being transformed. Ideals of quality of life that transcend the never-ending accumulation of material objects are now politically effective. The new politics of "participation" requires a broad sharing of knowledge, and therefore of skills and power. The philosophical perspective in which this argument is cast is one of the complementarity of "knowing-how" and "knowing-that", where uncertainty and quality are essential attributes of knowledge, and finally where there is a dialectical interaction of knowledge and ignorance.

CHAPTER 5

MEASUREMENTS

The results of measurement operations in the experimental natural sciences have long been assumed to provide a bedrock of certainty for human knowledge. Although every well reported experiment has its record of uncertainty, in its random error or inexactness, it is thought to be controlled through statistical analysis. There are other, deeper sorts of uncertainty, not familiar outside the matured natural sciences; their management is left to implicit craft skills or perhaps even neglected altogether. What cannot be clearly expressed will tend to be left in obscurity, in thought and in practice. For the management of those other sorts of uncertainty, researchers must deploy skills which are not usually associated with the natural sciences, but rather with those of history. We shall discuss those skills, and show how they are (quite unselfconsciously) applied in the formation of the judgements which govern the making of quantitative measurements and the assessments of their quality.

5.1. HISTORY IN SCIENCE

In the ordinary divison of studies in schools and universities, few subjects are further apart than history and experimental science. History proceeds from mainly literary records, to construct a story about human activities in the past. The historians organize their data in terms of some preconceived framework of assumptions about the historical process, in which certain aspects are emphasized (e.g. economics,nationalism, class-interest, personality, etc), and the others necessarily neglected. Historical explanations are then strongly dependent on the chosen framework. Data from the same sources can support very different interpretations. Progress in history occurs partly through scholars bringing new insights to largely old materials.

At first sight, the procedures of experimental natural science could not be more different. The continuous improvements in experimental techniques renders older results obsolete and irrelevant. Progress in science occurs through the replacement of old data and (normally) the refinement of existing theories. Although experimental data do not strictly entail theoretical conclusions, it is rare for opposing theories to invoke the same data. Instead of relying on a partly subjective interpretation of literary sources, science rests on objective, ususally quantitative reports on the behaviour of some aspect of the natural world.

The distinction between history and natural sciences is, of course, not absolute. Some natural sciences, as astronomy and geology, have a strong historical dimension. Some fields of history use statistical methods for establishing general trends; and scientific progress has a special branch of history devoted to its study. Also, there are some aspects of methodology which are shared between the two types of fields; thus, both proceed normally by the attempted assimilation of new data to existing theories and explanations. This is necessary for the continuity of the intellectual structures of the subject, persisting through their inevitable changes.

We argue here that the history of scientific endeavours is relevant, not merely at the general level of the succession of grand theories, but also in every particular case where an important quantitative result is derived through experiment or measurement. Relatively few people are aware of how much variability and uncertainty is involved in even the most basic of "physical constants". To make sense of this variability, the researchers develop their own historical perspective on the interrelation of theories, devices and people, in the attempt to create reliable quantitative information about the workings of the natural world. The fact of such repeated significant changes in physical constants makes it clear that the scientists are not merely neutral observers of an impersonal physical reality; but they are interacting with it, through their subjective commitments as well as through their objective instruments. They themselves are part of an historical process.

For a significant example of the workings of history within experimental science, we reproduce a graph showing successive recommended values of a fundamental physical constant, the "fine-structure constant", α^{-1} (Figure 1, p 4). By comparing successive values with the "error bars" of each (representing their inexactness), we notice that the change from one to the next frequently lies far outside its expected range, at least until the recommended values "settle down". Roughly speaking, this means that there is less that one chance in several hundred that the difference between two sets of experimental data (providing the two successive recommended values for α^{-1} are the results of random processes. Such a phenomenon of excess variation is quite familiar to experimental scientists; it is usually described as "systematic error", equivalent to an unreliability in any particular set of readings or results. Such drastic changes come partly from improvements in instrumentation and technique; but they also result from new theoretical perspectives, or insights, about the "same thing" that is the subject of measurement.

A recommended numerical value is thus the product of an historical process, involving skills and judgements. For it is based on reports of diverse experiments, separated not only in place, but also in time and circumstances. Those who recommend the value along with its error bars make a critical survey of the available literature, and use their judgements for assessing the relative quality of different experimental results (even deciding when to reject some altogether in spite of their scientific appearance). This task is one of the most demanding in science, requiring several highly developed skills. Some of these are similar to those of history, in the discernment of the quality of written sources, and in the nuanced expression of critical evaluations. Thus in a comprehensive review of papers on the evaluation of a particular constant, all those *not* worth consulting may be in a list prefaced by "See also" (Ravetz, 1971, 274n). In addition the reviewers must have mastery of the relevant skills of experimentation, so as to invoke "limits of the possible" in their evaluation of research reported, as done on particular equipment by particular people. The final recommended values as it appears is, in one sense, purely the product of history, for it need not coincide with any particular reported value, being a consensus achieved by selection and interpretation of sources.

It is not commonly appreciated, just how severe that selection and interpretation must sometimes be. A graph for experimental value for the thermal conductivity of copper reproduced by Tufte, shows a sequence of the recommended values for different temperatures to be those recorded in just one paper, while the others are all less than those, many by a factor of 10^2, or even 10^3. The discrepancy is explained by the presence of impurities in the samples used in the other laboratories (Tufte, 1983, 49).

The assignment of systematic error is quite clearly a matter of historically informed judgement. But even the random error accompanying the reported value of a constant may be the outcome of a complex train of reasoning (for a full discussion on the subject, see Henrion and Fischhoff, 1986, 795). Thus, some scientists

cautiously assign unreasonably large errors so that a later measurement will not prove their work to have been "incorrect". Others tend to underestimate the sources of systematic error in their experiments, perhaps because of an unconscious (or conscious) desire to have done the best experiments. Such variation in attitude, although out of keeping with scientific objectivity, is nevertheless unavoidable so long as scientists are also human beings. (B.N. Taylor *et al.*, 1969. Quoted in Henrion and Fischhoff, 1986.)

In terms of NUSAP, we can interpret the first of these strategies as a trade-off between the spread and assessment categories of the system, where an increase in the spread is used as a means for conveying less confidence in reliability.

History also enters into the social process of interpretation and common use of such constants. For everyone in the field knows that at the next periodic revision, the authoritative accepted value will be different from the one currently in use, perhaps even lying outside its error bar range. However, for the sake of consistency and continuity, researchers design and calibrate equipment around the existing values, fully aware of their possible "incorrectness" due to a large systematic error, pending the future version.

5.2. UNCERTAINTIES AT THE FOUNDATIONS OF SCIENCE

Such instabilities in the quantitative foundations of natural science do not usually damage its superstructures, for several reasons. First, in a matured natural science, the variability in fundamental constants is usually relatively small, perhaps a few parts per thousand or even less. When such numbers are used, either for calibrating equipment or directly in experiments, the instruments are set up so as to be robust with respect to uncertainties of the magnitude that have been encountered. Also, as the graph of the successive recommended values of α^{-1} shows, progress in a field eventually produces a "settling-down" of the constants so that the systematic error, as revealed in the differences of successive recommended values, is no greater than the random error of either. When a constant becomes very reliable in this sense, its theoretical interpretation is liable to be suddenly tranformed. For the "best" in an inter-connected set of fundamental measurements can be adopted as the new defining standard. Thus, for example, the "meter" changed from being a length on a platinum rod in Paris, and became a multiple of the wavelength of a particular sharply-defined spectral line of the element Krypton-86.

The process of assimilation of new results to old also enhances stability and reduces uncertainty; thus if a particular instrumental approach yields values which are drastically different from the consensus, the research community will scrutinize them with special rigour, searching for a flaw, The classic case here is the phenomenon of "ether-drift" recorded by a distinguished physicist in the 1920's, contrary to Einstein's theory; and it was eventually explained away. But it could be a mistake to believe that accepted scientific results are purely the outcome of "negotiation" leading to a consensus, as argued by those who advocate a "constructivist" theory of science (Brownstein, 1987). Even the more surprising results will be accepted when they pass scrutiny for their validity and fruitfulness. The most recent famous case in point is the discovery that inside the earth the force of gravity is some 20 % higher than in free space; this is a systematic error some thousands of times larger than the accepted random error.

The example of the uncertainty associated with the measurement of fundamental physical constants illustrates the thesis that the process of quantitative measurement is not at all a simple, straightforward operation for obtaining a number which describes a physical object or a situation. To put it in a positive way, we may say that every measurement is the result of a complex set of operations, in which skills and judgements are deployed, and which possesses an historical dimension in several respects. The art of measurement in an experimental science consists largely in the management of the characteristic uncerainties; it is in that way that the information is controlled and assured. In these fields, the skills of good experimentation are passed on by example and precept, from mature scientists to the students. They are learned by imitation, trial-and-error and informal reflection.

Such craft skills are surprisingly similar in principle to those in traditional handicraftsmen, and are not easily transplanted or created afresh in new fields of inquiry. In particular, in policy-related research, the acute problems of quality-control which we described in a previous chapter are in part a consequence of their novelty, and of the absence of traditions of craftsmanship in the management of uncertainty and the assurance of quality. In such circumstances, the craftmen's approach will need to be supplemented. Tools and techniques whereby the existing skills can be guided and extended to good effect are urgently needed. Our approach is to generalize from the successful practice of the matured natural sciences, without falling in the trap of a reductionism that treats all sciences as imitations of physics. We are providing a general codified framework, whereby skills for the management of uncertainty and for the judgements of quality, appropriate to these new fields, can be mastered and deployed.

We will show that no physical measurement can be a completely exact and reliable description of a quantitative attribute of some physical system. When, for example, we say "length", we are making reference to a complex, conventional and theory-laden process which produces a number to be associated with an aspect of a physical body. There are unexpected skills to be deployed even in the measurement of the length of a rod by a meter-stick. Our analysis of measurement runs parallel to that of the previous chapter, where we showed that an arithmetical language cannot be completely formalized, neither within itself nor in its appplication to science. Here we show that the objects of that language, including measurements, experimental results and other numerical constructs, have uncertainties which cannot be eliminated but whose quality is assured by their effective management. In a later chapter, we will use the analogy between numerical systems and maps, to show that any quantitative description is designed in accordance with certain criteria of quality, with its eventual functions in mind. On that basis we can then establish the relevance and importance of a new approach to notational systems for the expression and communication of quantitative information.

5.3. N.R. CAMPBELL: MEASURING LENGTH

For our argument, we go back to the original analysis of measurements in physics by N.R. Campbell (Campbell, 1920). Much work has been done since then, mainly motivated by the problems of quantitative psychology, particularly the construction of different kinds of scales (see, for example, Stevens, 1946). However, none of this later work addresses the fundamental issue, namely the actual relation between formal systems and operations on the material world. This nearly complete neglect of a work whose importance is practical as much as philosophical, calls for an explanation. B. Ellis, in a standard text on the philosophy of measurement, provides the following,

One reason for this neglect may be simply that there no traditional philosophical problems of measurement to challenge us... There is thus a climate of philosophical complacency about measurement. It is felt that nothing much remains to be said about it, and that the only problems that remain are peripheral ones. It is recognized that there are some unanswered questions concerning microphysical measurement, and some difficult problems of formalization (for example in constructing axiom systems for fundamental measurement). But for the most part there is little dispute; and hence, it is felt, little room for dispute. Much of the fundamental work of Campbell, for example, has never been seriously challenged. But why should there be this climate of agreement? One can only believe that the agreement is superficial, resulting not from analysis, but from the lack of it. Philosophers ranging in viewpoint from positivists to naive realists all seem to agree about measurement; but only because they have failed to follow out the consequences of their various positions. In fact, I believe the positivists are largely to blame for this situation. For usually they have proceeded from a concealed realist standpoint. It is not surprising, therefore, that realists should find little to argue about. (Ellis, 1966, 1–2.)

Campbell started as a physicist. He discovered basic problems involved in the operation of measurement, during his work on setting standard units for electrical measurement, as the volt, ampere and ohm. He discovered that the skills of ordinary laboratory research do not suffice for competent work on these kind of subjects. Clear and consistent thinking on fundamental measurements requires awareness of the uncertainties introduced by theoretical assumptions and variable calibration constants, together with the logical structure of the operations. In his first major work he attempted to show how, in principle, all of physical measurements could be built up from those of the most simple and theory-free sort. By the end of his book he recognized the impossibility of that ideal. In his later work (which we do not discuss here) he attempted a classification of the different sorts of measurement as they occur in the practice of physics.

We shall now adapt Campbell's analysis to some simple, familiar measurements, which are commonly thought to provide a "true" result or the nearest thing to it. We take for example the case of the measurement of length of rods, on the scale of parts of a meter. The operation involves putting a meter-stick against the rod to be measured, and reading-off the number which corresponds to the relevant gradation, perhaps with an extra digit for estimation of the space between the finest gradations. The first path to a deeper understanding of the process is opened by the question, is one reading sufficient? If not, why not? This is easily answered, for we all know that for a truly scientific result, there must be a series of readings, since a single reading is liable to involve some kind of error. So then we ask, how many readings should we take? A decision must be made. The answer will depend on the required quality of the result; this is defined by its accuracy on the one hand, and its cost and feasibility on the other. Clearly in almost all cases there is no need for a protracted analysis of this design problem; "a few" readings in this simple situation are enough. But the philosophical argument is not so easily settled; and we ask, by what criteria do we decide the number of readings? These depend on theoretical assumptions on the behaviour of the proposed set of readings as a statistical ensemble. If they are "normal" or "near enough" to that, then certain mathematical manipulations (applying the theory of errors) can be deployed to relate the desired inexactness to the required tries. But if the expected distribution is not near enough to normal, then some other theories, with their appropriate mathematics, must be deployed. How do we know that the expected distribution will be "near enough" to normal? There are practical solutions to this question; partly the history of similar operations, and partly some higher-level theories about the relations of empirical and theoretical distributions. But rather as in the case of the recursion of metalanguages, the sequence must become informal at some point if we are to avoid an infinite regress, which in this case means paralysis. Thus the simple question of "how many trials?" opens up a Zeno-paradox situation of necessarily prior decisions, which can be broken only by craft skills guided by judgements.

However, there is yet more to come! Suppose that we have decided on the numbers of readings; we have done them, and we have a set of numbers representing the measurements. What do these numbers tell us about the true length of a rod? Clearly, that is most likely to be between the extreme values we have recorded. But which, if any of the intermediate values, is the true value? To determine this, the usual procedure is to perform some arithmetical operations on the numbers and obtain a representative, which is usually the mean or average. Two questions remain: first, what relation does this representative have to the true value; and, second, why choose this rather than some other representative of the set, as for example, the median or the mode? For the first question, the answer is provided by higher-level theories about the relations between estimates and the true value; as we could now expect, these depend on empirical assumptions whose validity is decided by judgements. If this solution is theoretical, the answer to the second question, about the choice of representative, can have very practical consequences. If the distribution of results is significantly different from normal, and perhaps is skewed or has outliers, then that mean can be a positively misleading representative, and certainly not the true value. (Social statistics are particularly useful for showing how particular representatives can be misleading. Thus, "average" income in a village with ten poor peasants and one rich landlord is not well represented by the arithmetical mean.) In cases of non-normal or non-symmetric distributions of data, there must be a decision or choice of representative, based on craft skills guided by judgements.

The outcome of this line of questioning is that the measurement of "the length of a rod", a straightforward operation in ordinary practice, conceals within itself a host of assumptions, theoretical and empirical. The conclusion (which Campbell reached so reluctantly) is that we can in no way measure "the length of a rod"; we can at best perform a set of measuring operations, which depend on both mathematical theory and craft skills, and hope on the basis of past experience that they are adequate.

More problems arise when we wish to combine lengths, to apply arithmetic to them for comparing or adding. Without such operations, there is no possibility of science. It is well known that for a system to be capable of arithmetical representation, it must satisfy certain properties, expressed formally as axioms. Among the most basic of these is the transitivity of the equality relation; in symbolic form: If $A = B$ and $B = C$, then $A = C$. It may be surprising to discover that measured length does not satisfy this axiom. We can see why, by analysing the meaning of equality in this context. Equality here is not the identity of pure mathematics, for every measured quantity has an inexactness or tolerance. Whether two of these quantities are to be deemed equal is again a question of the sort we already discussed above. It is unrealistic to demand that their means be arithmetically identical: for then we would certainly deem pairs of rods to be unequal when good sense dictated otherwise. So we adopt some additional rules; for example, two rods are equal or equivalent if they differ by less than some preassigned amount or percentage. This takes us to another Zeno-type paradox, this one less easily escaped. For suppose rod A and rod B differ by two-thirds of the accepted tolerance, and rod B and rod C similarly; then rod A and rod C may differ by more than the established limit. So then $A = B$ and $B = C$, but $A \neq C$. Thus the relation of equality, possessing transivity as one of its defining properties, is *not* transitive in this instance. In other words, the quantities obtained by measurement are *not* numbers as commonly understood, because they fail to satisfy this essential property of numbers. We shall later discuss how this divergence between measured quantities and numbers has been rendered harmless in practice, and so is capable of being safely ignored.

Such a paradoxical result is already familiar in some fields, as of various sorts of subjective perceptions. For example, people can report two pairs of colours as identical, and yet see the difference between the extremes. This phenomenon may be taken as evidence of the imprecision and subjectivity of that sort of perception; but here we find it applied to the measurement of length, and hence in principle to any kind of physical measurement. The conclusion of all this is that basic physical measurements are not characterized by rigid logical structures whereby we can describe aspects of reality without recourse to craft skills guided by judgements. We have seen how the assignment of numerical values to lengths, and the decision about the equality of lengths, exemplifies this conclusion.

Finally, we will show how the very operations by which this simple length is measured, embody theoretical assumptions about the thing itself. The point can be established by the question, "how is length measured?". A student might simply rest one end of the rod and an end of the meter stick together against a perpendicular surface, and then read off the gradations of the meter stick where the rod ends. The paradox is that this is a mistake, for it is based on the assumption that length is a function of a single point. In fact, length is an interval-measure, and so it depends on the position of two points; and in a proper measurement the tolerance of each of them should be accounted for. This is the reason why some experienced physics teachers rather pedantically insist that the starting end of the rod be placed at some point other than zero. In practical terms, attention to this detail doubles the inexactness of any recorded measurement of length. In principle it reminds us how easy it is to be mistaken about the appropriate measuring procedure for any physical attribute.

5.4. TEMPERATURE: MEASUREMENT AND CALCULATION

Conceptual analyses of science harbour a special pedagogical difficulty: if the examples are taken from active research science, they are incomprehensible to all but a few highly trained researchers; but if they are elementary, they have an air of triviality. We shall therefore consider an illustration of an inter-mediate class: temperature. This is a property of bodies that is familiar in ordinary life; and it is also a basic concept in physical sciences. It might appear that the measurement of temperature is no different in principle from that of length: the distance along a scale of an expanded liquid is normally the thing to be observed. A closer examination of the theory and practice of tempera-ture shows how very much more complex it is, than length. The measurement itself involves significant disturbance of the system being measured; for the "probe" which senses the temperature cannot be identical to the body or space being measured. Although this effect is unimportant in ordinary clinical con-ditions, in refined physical measurements it must be reckoned with. It brings to mind a principle of intrusion first enunciated in connection with anthro-pology. This is, that there can be no neutral observer; every act of observation is in fact an interaction with the system in study. Quantum physics is a familiar example of scientific activity where this principle is in operation; but there it is assumed to be significant only at the micro-level. It is also quite obviously present in the life sciences, where researchers talk about intrusive or (rela-tively) non-intrusive methods.

Temperature as measured generates the same problems in relation to arithmetic as those discussed in connection with length. There are added difficulties, more typical of certain theoretically-based concepts. In the ele-mentary study of heat, students normally take an average of temperatures. What can this mean, physically speaking? Arithmetically, an average is one-half the sum, resulting from the addition of the two quantities. Adding two lengths is straightforward enough: one normally puts the rods end-to-end and observes that the length of the joint rod equals the sum of the lengths. In this case, how are we going to add physical temperatures? Only by increasing the heat content of a body, to achieve a new temperature equal to the given sum. But this is never done; it does not correspond to the elementary operations in the mixing of bodies for calorimetry; and in any event is usually not feasible experimentally. Hence the arithmetical operation of addition does not have a direct physical correlate. This is equivalent to saying that temperature is an intensive magnitude, like density (mass per unit volume; how could samples of these be added?), as distinct from the extensive magnitudes such as mass and length. These distinctions are themselves far from simple (for an ex-haustive discussion, see Campbell,1928). Interestingly, the operation of averaging can have physical meaning for other intensive magnitudes as well. Thus, just as we mix fluids to get an average (possibly weighted) of their temperatures, we can do the same to get an average of their densities. Hence we may say that whereas the arithmetical operation $(A + B)$ has no physical meaning, the apparently more complex one $(a \times A + b \times B)/(a + b)$ does! Thus ordinary arithmetic is applicable only in part.

5.5. UNCERTAINTIES IN PRACTICE AND THEORY

These examples show how the simplest arithmetical operations can be problematic when applied to elementary physical measurements. What happens in practice, so that the formal systems can function effectively, is the unselfconscious development of routines and craft skills to avoid the pitfalls that would otherwise be encountered. The applicable axioms of the formal systems of measurements can then be used to guide the development of appropriate skills. Thus the non-transitivity of the equality relation is rendered harmless by the use of a single standard for measuring a set of copies. In any relevant field of practice, it is understood that to compare copies against copies successively is to risk a "drift". Put otherwise, such a maxim of practice ensures that the conditions of applicability of the formal system are effectively maintained. Good craft practice thus includes rules (whose rationale may be unknown to practitioners) for this function. Thus the link between "pure" mathematics and "applied" mathematics is established only partly by formal rules of interpretation; these are necessarily supplemented by skills and judgements.

There are other, more advanced examples in science which illustrate this point. Thus the common arithmetical operations of multiplication and division, applied to physical magnitudes, presuppose theories of units and dimensions which are highly articulated and themselves not free from paradoxes (see, for example, Bridgman, 1931). When we go beyond these measurements which at least seem elementary, new sorts of uncertainty are encountered. Any sophisticated scientific equipment can and will malfunction. There is the aphorism, known humorously as the "Fourth Law of Thermodynamics": no experimental apparatus works the first time. Insights such as this one are the stuff of the "Murphy's Law" literature. The philosophical point they express is that any deliberate action is based on hypotheses (usually implicit and even unselfconscious) concerning the structure and relevant properties of the reality being affected. Instruments, as is so well described in the "Murphy's Law" literature, have inherent unreliabilities in their operation. Components which are ill-matched or unreliable in themselves, eventually cause the collapse or destruction of the systems in which they are embedded. The fact that Chernobyl and Challenger are (as yet) the exceptions, shows that quality assurance in the technological sphere is still generally effective. How are these reliabilities controlled? If we simply trusted that all our tools and instruments were generally good enough, then our high-technology civilization would soon grind to a halt amidst widespread breakdowns and disasters.

There is no perfect solution to the problem of unreliability; its practical resolution is accomplished by good working practice. An important element of this is calibration. This is a more sophisticated form of measurement; it is the process whereby the response of an instrument to its inputs is obtained by a disciplined method of enquiry. Calibration also achieves the separation of the different sorts of uncertainty. Thus the "bias" (or systematic error) can be identified when there is a standard available for comparison, and then corrected. Then the imprecision, due to the combined effect of the small uncontrolled causes of error, can be studied separately. A calibration process depends on the assumption of constant values for all inputs except the one under study. Since there is always some inexactness there, it must be controlled as by the outputs being insensitive to inexactness of the inputs at the expected size. Similarly, when a standard is involved, it must have much less inexactness than the instrument, lest random and systematic errors be confused in the measurement. For the standard itself, the question of its reliability in relation to its use may be crucial, and hence an evaluative history of it can be important. Many instruments are complex in their structure and operation; and so simple calibration of their constituents one by one does not suffice for the full assessment of reliability; highly developed craft skills are then needed for calibration; and skills of expression are needed for conveying its results. Indeed it is in connection with calibration that we see most clearly the need for a system in which the various sorts of uncertainty are distinguished. (For a full description of the process of calibration, see for example, Doebelin, 1986.)

Further, the design of scientific instruments is based on theories of their operation, and these relate to theories or models of the process under scrutiny, which necessarily simplify the realities involved. These uncertainties may produce a deeper sort of systematic error: that which is being measured may be an "artefact" of the instrument, rather than an aspect of the system under study. It is not easy to provide familiar historical examples of such problems, since by their nature they do not occur in the big, successful science that is recorded in textbooks. The variability in physical constants, which we discussed above, is partly explained in terms of this unreliability effect. Only occasionally is such a phenomenon worthy of historical attention. Among the most familiar is the noticeable secular variation in "Hubble's constant", which measures the rate of expansion on the universe and thereby its age.

Beyond the unreliability of instruments, we have the uncertainty as to whether the thing being measured (or studied) exists. The discovery that such a "thing" hitherto accepted by science, is a "no-thing", is naturally a cause for concern, philosophical and practical. It may sometimes be the occasion of a "scientific revolution", as was the rejection of the supposed motions of the sun and stars on the traditional astronomical system. Or it may be a less dramatic reconstruction, as when Kepler's ellipses replaced the fictitious uniform circular motions that were accepted and measured by Ptolomy and Copernicus alike. Even quite sophisticated physical theories may be discovered retrospectively to be about non-existing entities. A relic of such a theory survives in elementary physics: "calorimetry" is the measurement of "caloric". This was a highly plausible, and indeed scientifically effecive theoretical entity (Carnot used it in his creation of thermodynamics) which was eventually discarded. Even more highly elaborated was the "luminiferous aether" of the later nineteenth century. This still survives in ordinary language, in connection with radio communication; and it is indeed difficult to imagine the electromagnetic waves in the absence of an undulating substance. The most prestigious scientists of the later Victorian age tried to construct mathematical or mechanical models with the requisite properties; and their endeavours were rendered meaningless by the success of Einstein's theory of relativity.

5.6. SCIENTIFIC UNCERTAINTY: PHILOSOPHY AND PRACTICE

We have shown how even in the traditional sciences, measurement is affected by characteristic uncertainties at all levels, from the technical to the ontological. In policy-related science, no research is immune to the occurrence of such uncertainties, frequently in a very severe form. The concept "risk", which is so central in many environmental and technological issues, involves many uncertainties in its quantification. To begin with, "risk" is commonly taken to refer to some unwanted event; its measure is a function (usually the product) of its likelihood and its harm. Measuring this likelihood, as we mentioned in Chapter 1, an inevitably inexact operation. The events themselves may not yet have occurred, or may be extremely rare; relevant data may be very hard to come by. In many cases, empirical data are so deficient that risk assessments are based on computer models or expert opinions. But, as Beck has shown, these are not really part of scientific methodology; and so the uncertainties associated with their outputs, are not mere inexactness, but unreliability. The same holds even more strongly of the measurement of "harm", which depends crucially on conventions and value-laden assumptions (what is the cost, or value, of a life or a limb?). Underlying all these uncertainties are the problems of causation; what makes such events occur rarely, rather than frequently or never? There are many different kinds of causal links connecting initiating events with accidents, and accidents with resulting harm. There will be always an interaction between the "hardware" and the "software" (including monitoring systems). The different models for "major hazards" are quite contradictory in their assumptions and implications for practice and policy. Broadly speaking, we ask whether occurrences are capable of probabilistic description, the acts of a dice-throwing God; or whether they are man-made, the foreseeable results of inadequate monitoring and morale (Turner, 1978). Hence the issue of what risks *are* and whether they can be measured, remain unresolved, the subject of protracted debates.

The examples in this chapter have taken us from technical uncertainty, as examplified by random error, through the methodological uncertainty of systematic error, finally to the epistemological uncertainty of whether our scientific theories relate to the real world. At this point we find ourselves addressing issues which belong more to the realm of philosophy than of science; for there is no conceivable test for resolving such problems. Within the received view of the philosophy of science, problems are studied because of their relevance to the philosophy of knowledge. Such studies usually come in the form of abstract logical reconstructions, sometimes ignoring or even contradicting the history and practice of science. All the tendencies within this general philosophical paradigm share one common feature: the absence of any recognition of the practical uncertainties of scientific research. It is now widely recognized that no scientific results are immutable; indeed for Popper falsifiability is the criterion of real science, and willingness to admit error defines the true scientist (Popper, 1935). But Popper's scientific hero was Einstein, who was rewarded for his integrity by being right. He never discusses the occurrence of errors in the history of science, either how great scientists sometimes committed them or how they coped when they were discovered. Thus Popper's uncertainty is of a purely metaphysical character; it has nothing to tell the workng researcher. For this whole school of the philosophy of science, the practice of science is like flickering shadows on the walls of Plato's cave in the *Republic*, a feeble caricature of its rational reconstruction.

Contrasted to this dominant paradigm in the philosophy of science, there was the tradition of reflective scientists, working at the creative boundaries of their fields, at the limits of scientific feasibility, who attempted to provide a philosophical (and sometimes historical) perspective to the problems of a deeper understanding of their practice. The pioneering work of E. Mach (Mach, 1883) was followed by P.M.M. Duhem (Duhem, 1914), N.R. Campbell and P.W. Bridgman (Bridgman, 1927). Such a concern was related to the leading problems of physics in the later nineteenth century, when fundamental measurements of all sorts, and in particular those of the rapidly developing electro-technology, were seen as crucial for practice. In England, the threat of German supremacy in this field was recognized; and Imperial College in London and the Natural Science Tripos at Cambridge were two responses at the highest academic level. Campbell had his scientific formation in this field, where the problems of coherence among instruments and standards were more challenging and real than those of the inevitably hypothetical general theories, as of the "luminuferous ether" (Warwick, 1989).

But quite soon, such problems became merely technical, relegated to the National Physical Laboratory; and in atomic physics, leading experiments could be done in the "sealing wax and string" style of the Cavendish Laboratory under Rutherford. The interests of the philosophically-minded physicist was diverted to the more abstract problems raised by Einstein's work in quantum theory and relativity. In those debates, some key experiments could start life as "thought experiments", becoming technically feasible only much later. This sort of philosophical reflection related naturally to mainstream philosophy, extending even to ethics and theology. In contrast, the concerns of the "reflective working scientist" tradition could too easily be dismissed as specialized craftman's wisdom, occasionally supplemented (as in the case of Bridgman) by amateur philosophy. In its concentration on the analysis of practice, Campbell's reflections on the foundations of measurement do not even read like "philosophy", while his more popular book, *What is Science?* (Campbell, 1921), much more conventional in its contents, can pass.

In a different tradition and addressing a different "problematique", the French philosopher G. Bachelard analyzed the various obstacles to the achievement of scientific knowledge (Bachelard, 1938). His study was one of the few to go beyond Francis Bacon's "Four Idols" as the sources of ignorance and error. Among the obstacles Bachelard discussed were those relating to quantitative knowledge; and he provided examples of pseudo-precision among the great men of science of the past, as Buffon. Unfortunately, Bachelard's name is best known through the term "epistemological rupture", which was appropriated by Louis Althusser. His contributions to the philosophy of science had very little impact on English language studies in the field (Tiles, 1984, xi).

With the advent of policy-related research, measurement has again come to the fore. In one sense the problems are completely different, for the crucial concern is not with highly precise measurement of tightly controlled phenomena. Now, the instruments, however sophisticated technically, are used in conditions of severe, even extreme uncertainty. To distinguish between artefacts and actual effects, or to determine "representative samples", to say nothing of establishing statistical significance of data or tests, are the major tasks. And there is always the policy, political dimension, influencing not merely the reception of the results but the framing, and indeed the existence, of the research project. Thus policy-related research raises so many urgent and inter-related problems of method and practice, that this strand of philosophy of science again becomes the most fruitful of enquires. By using the NUSAP system to distinguish among the different sorts of uncertainty, we establish links through from the technical aspects of inexactness through the methodological problems of unreliability, to the epistemological issues of border with ignorance. Such distinctions, enriched and clarified in our explanations and applications of NUSAP, may also provide new conceptual tools. By their means the craft wisdom of the reflective researchers may be expressed and organized in a philosophically coherent way, so that enquiries recognizable as philosophy can be conducted on the basis of their insights. This will be a philosophy arising from the practice of policy-related research, and devoted to its improvement.

CHAPTER 6

MAPS

Maps provide an interesting and significant contrast to the sorts of quantitative information that we have discussed hitherto. First, they are not affected by the "magic number" syndrome; although they can be used to make strong or contentious claims, they do not possess the aura of objective truth in the same way as numbers. Because of this cultural difference, maps are now generally accepted as being the product of human creation, embodying policies, prejudices and error. The imperfections in maps are therefore not an occasion for dismay; nor do philosophers need to argue that somehow they still belong in a Platonic heaven of quantitative science. Modifications in maps are generally appreciated as being the result of changes in both scientific knowledge and political realities: philosophers have not needed to articulate general theories of "cartographical falsifiability" or "cartographical revolutions".

Maps, as the obvious product of design, are particularly useful for our analysis of quantitative information. It is easy to see the influence of design choices on maps, by comparing two maps on the same scale and the same function (as for example, motoring maps), but produced by different publishers. Further, maps have an intimate relation to uncertainty, which is shown clearly in their conventionalized features. Someone wanting to know the extent and shape of a smaller town or village on the ground, can not learn it from the dot, cicle or disk on the small scale map. Some information is simply omitted from some maps, while represented on others at the same scale.

The skill of the mapmaker is involved in helping the user avoid the pitfalls of misinterpretation, by design which makes it clear which sorts of information are provided with a stronger claim of certainty, and which sorts are less certain or ignored. The different sorts of uncertainty, as we have analyzed them, are easily distinguished in the design of maps; and this feature will be useful when we establish the analogy between design in maps and graphics, and design in numerical systems. The assessment of quality in maps, involving adequacy of representation in relation to objects and function, will also provide us with a useful analogy for numerical systems.

6.1. "SOFT" MAPS V. "HARD" NUMBERS

Superficially, maps are very different object from numbers. The one presents a large amount of information in a graphical form, which is inherently imprecise. The other has a single item, in a representation that enforces precision even in its statements of inexactness. The map emphasizes the totality; the individual elements are seen and grasped in relational terms. By contrast, the number appears isolated (although any individual digit has meaning only as part of a succession). Phenomenologically, the map is *Yin*, soft, a matter of fields and suggestions; while the number is *Yang*, hard, consisting of atoms and assertions. By its Gestalt, the map expresses vagueness, and encompasses uncertainty; the number is unambiguous and precise. The understandings conveyed by a map are partly implicit, as in a Wittgenstenian family of meanings; the number's message is direct and simple. One can browse over a map, dreamily, for hours; with a number the experience is short and sharp. It is no wonder that numbers still produce metaphysical Angst by their failure to deliver the certainty that their form promises; while maps, doubtless the subject of Platonic aspiration at some point in their history, are now accepted as tools.

Complementary to these oppositions of appearance, are the analogies of structure. Maps are, after all, created by a multitude of quantitative operations. Indeed, maps are increasingly produced in digital form so that their qualitative graphics are only the print-out of a data-base. Such data-bases may be the computer transformations of direct observations, as by remote sensing; the resulting maps and statistical summaries now may regain a scientific status similar to that of numbers. In this way the quantified outputs of modern cartography can become inputs to the policy process, hardly different from the data of experiments or field studies. In general, maps have historically reflected the progress of science; their production has become more mechanical, abstract and theory-laden, replacing the individual craftsmen with their simple instruments, refined skills and partly aesthetic judgements. Thus, they have become increasingly certain as sources of knowledge and bases for predictions. Paradoxically, in a way analogous to scientific theories, they have gained in apparent certainty of knowledge as they have lost in immediacy of experience.

Because of the dialectical relation between maps and numbers, a study of the properties of maps can illuminate those of numbers. Our basic point is that in both cases the representation is a product of design. In common with all other such products, this involves a harmonization of a set of discrete criteria of quality, including function or use. In the present case, where both are concerned with information, the design process includes the management of uncertainty as well. Whereas there is a traditional ideal that scientific knowledge should be an ever-improving approximation to reality, map making has generally accepted that, given its inevitable reduction of scale, useful representation is the most it can accomplish. Thus Lewis Carroll could make one of his characteristic jokes about a pedant ("Mein Herr") who believed that the perfect national map would be one on the scale of one-mile to the mile (Carroll, 1893, 556). (Unfortunately, the farmers objected when it was laid out flat). On any other scale, simplifications and conventional features are inevitable.

Naively viewed, conventional features could be considered as distortions or fictions. Thus invisible political boundaries are marked, conventional colouring is used freely, inhabited places are represented by circles; and even on large-scale maps (as 1 : 50,000), roads are drawn on a much larger scale so as to be visible, and representations of the buildings around them are therefore displaced significantly. Someone from a different culture, or perhaps just unfamiliar with cartographic conventions, could be misled by such deviations from verisimilitude; and it *can* be confusing when some conventions are "realistic" (for example, blue for the sea) and other not at all (for example, red for the old British Empire). In spite of these clear departures from verisimilitude, no educated user will dismiss maps as false or "subjective" on account of such features. The ideal of photographic perfection is simply not there. In the case of numbers, ever since Pythagoras some sort of metaphysical identity of representation and the represented has indeed been the ideal; hence the "magic number" syndrome that persists to this day. The issue now is how numbers are best to be demystified; for the situation in policy-related research is such that the dream of approaching reality ever more closely by science is not merely impossible but also now quite counterproductive. "Magic numbers" in such fields as risk assessment can only increase confusion and impede effective work. The analogy with maps, sophisticated and useful representations in spite of their obvious inherent limitations, will help us in our comprehension of numbers as they are best used in science today.

6.2. MAPS AND THEIR UNCERTAINTIES

Our analysis of maps as products of design starts by considering the various sorts of uncertainty that they convey. These correspond well with the categories of the notational system NUSAP. Every map has a characteristic "inexactness", so that one cannot determine the location of a portrayed object below a certain minimum distance, or even ensure the existence of an object of below a certain minimum size. These lower limits may be considered as defining the "grid", determined partly by scale but also by the permitted density of information on the map surface (thus a motoring map is simpler than a topographic map to the same scale). This is analogous to "scale" and "resolution" in remote-sensing photographic displays. This grid amounts to a topology on the "mapping" from the real, visible surface to its representation. The appropriate coarseness of the topology will depend on the intended function; a general purpose map can be very coarse, providing useful non-critical information at low cost in resources and therefore at a low price for the user. For inputs to decisions with greater sensitivity to error-costs or with higher costs or higher stakes, the coarse general-purpose map would be inappropriate (not "false"!) and a competent user would know to seek out more expensive, specialized information from the appropriate map. Thus maps can be considered as tools, with a variety of designs conforming to a variety of functions; these can be accomplished with greater of less effectiveness.

Another familiar basic design feature of maps (most noticeable on those covering very large geographic areas) is projection. Here the uncertainty is not of simple inexactness, but that induced by the distortions resulting from any given projection. Again, it is not a question of "falsity", since it is mathematically impossible to make a non-distorting representation of a sphere on a plane. So we can consider this effect as a sort of "unreliability" in that the users of maps must possess skills of interpretation lest they be misled. How many of us still think of Greenland as having an area as large as South America! Given the constants of human psychology, the location and shape of places on a graphical representation has great importance for our image of the actual world. Antarctica at the centre of the picture makes us feel rather odd. More to the present point, an analysis of the Northern circumpolar regions would, from such a map, be confusing and ineffective. Such choices, depending on history and politics rather on the inherent constraints of cartography, show clearly how choice of projection is a design exercise. Historically the different geometrical rules for constructing the projections (notably Mercator's) were explicitly designed around particular technical functions, as for navigation, at the price of their unreliability in other aspects. The many map projections now in use would in themselves provide an excellent object lesson in design.

Reliability, or its opposite, is also achieved by a variety of special techniques. One of the most interesting is "political cartography", where national boundaries are defined in accordance with national policies; some political entities are either re-named or deemed non-existent. Thus for many years the territory of the German Democratic Republic (D.D.R.) was labelled on many maps as "Under Russian occupation" and much of Western Poland as "Under Polish occupation". Even local maps may practice economy with truth in detail; thus on recent issues of the British Ordnance Survey maps, some military installations are obliterated, fictitious field-boundaries and streams taking their place. In the opposite direction, maps in less-developed regions may portray aspirations rather than engineering realities, in the form of planned (or desired) transportation links (motorways, bridges, etc). Thus in the case of maps, as indeed of numbers as well, the user should remember the maxim: reliability goes down as political or economic sensitivity goes up!

These examples of policy-driven unreliability should not cloud the issue of the inherent inescapable limits of applicability of any map. Data-collection and processing cannot always provide fully comprehensive and continous updates. A map of any region subject to change may be considered as similar to the case of the physical constant discussed previously. Then we saw how its recommended value is the product of an ongoing procedure of historical review, so that in between such reviews it is technically obsolete.

In general maps suffer in comparison with basic scientific information in one important respect: quality assurance. Perhaps because they are more in the "software" realm than the "hardware", their producers and publishers (with some honourable exception) do not feel it necessary to provide any "pedigree" for their editions. Some popular local maps do not even carry a date, only saying "latest edition" through all revisions! We will later discuss the question whether quality matters, particularly for maps; it seems that in practice, on quality assurance, the answer here is the same as that for almost all information outside the narrowly technical sphere.

6.3. THE "BORDER WITH IGNORANCE"

Maps may fail to provide a faithful representation of their realities in yet other ways. The history of cartography conveys graphically a concept of which we make extensive use in the NUSAP notational system: "the border with ignorance". Maps of previous centuries clearly display their coarse topology or inexactness (through scale, etc), and their unreliability (through mis-locations or distortions of grid, which are now obvious). Their borders with ignorance were also explicitly represented, by the blank parts of the map; but they were also, most dramatically, conveyed by the imaginary creatures therein, representing the hopes and fears of makers and users (Fig.2). Such aspects as these are not obvious on modern general-purpose maps in devel-oped countries; these after all derive from strong traditions of scientific cartography. But if we go either to less-developed countries or to new kinds of special-purpose maps, then all the three sorts of uncertainty which qualify technical information (inexactness, unreliability and border with ignorance) are relevant.

In the case of numbers (as well as in scientific theories) the border with ignorance is usually concealed. There is no analogue to the blank spaces or

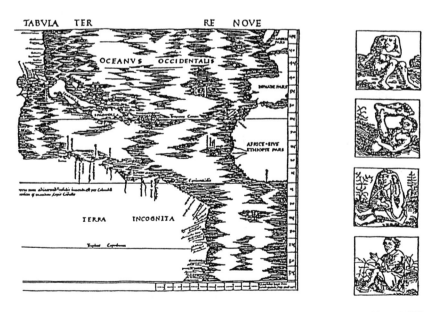

Fig. 2. Renaissance maps: blank spaces and monsters (from Bagrow, 1964, 108 and 199).

mythical creatures to indicate the presence or possibility of ignorance. The procedure and ideology of Kuhnian "normal science" (Kuhn, 1962), particularly strong in the teaching context, induce an illusory feeling of certainty in the permanence and truth of numerical facts and of scientific knowledge. Of course, scientists will always proclaim the tentative and open-ended character of their knowledge; but in the absence of effective techniques for analyzing and expressing the deeper uncertainties of science, these statements remain as pious protestations with very little influence on the otherwise dogmatic processes of the intellectual formation of students.

In practice, competent researchers know and cope with the various imperfections of their materials, cognitive and physical. they will have informal, perhaps tacit "pedigrees" for anything whose reliability is important for their activity. The blank spaces in their intellectual maps are of critical importance, for there may lie the most dangerous pitfalls. A formal pedigree, as developed for the NUSAP system, corresponds to the border between the detailed and the blank portions on the old maps, and stresses the presence of such blanks. Paradoxically, contemporary maps have lost all indications of the areas of ignorance. For obvious reasons, there are few blank spaces; and as we have mentioned, the user of commercial maps and atlases is hardly ever given any indication of unreliability, or any historical information about their provenance. By contrast, official maps, as the British Ordnance Survey, at least indicate the revision sequence, and accompanying publications explain the technical background.

The blanks in the old maps provide us with an example to illustrate a philosophical point very relevant to practice. When the mapmaker leaves a blank space, he is advertising his ignorance. The user of the map is made aware of the absence of knowledge concerning that place. Were there to be a merely empty place, like so many smaller ones on any map, the users would not necessarily be alerted. They would remain in a state of "ignorance of ignorance". The pitfalls of those areas of compounded ignorance are even more dangerous, for there is no signal that they might even exist. Thus in such practical cases, ignorance is not a mere absence of knowledge; it is a condition with direct consequences; and it must be "talked about", lest practice suffer. This might appear to be a contradiction in terms, for ignorance, by definition, is beyond analysis. But it is possible to speak about the border with ignorance; and this is a function of the pedigree in the NUSAP system. By its means, we can signal what has *not* been achieved in the production of an item of scientific information. Users are thereby protected from assuming a strength in the information that its "scientific" appearance may erroneously suggest.

6.4. MAPS: WHY QUALITY COUNTS

There is a final topic to consider, in relation to traditional maps. This is whether quality really matters. The same question can be raised, of course, about quantitative information. In both cases, almost all instances of use require a very low degree of precision. Do we really need such fine detail in maps; and also does the uncertainty of information matter outside a few special areas? We recall that quality of information involves function (including cost) along with production (including uncertainty management and correspondence to its object). Hence for the casual uses of a map, just as for the publication of a number in a political context (to say nothing of quiz-game), the criteria of quality are undemanding. In the case of a map, simply to confirm that Milan is further north than Florence does not require a very fine topology or a recent critical revision. Similarly, the height of Mt. Everest, given to the nearest meter, needs only to be consistent between the sources of the quiz-show contestants and their judges. Indeed, it is likely that most of the detail in most published maps is if a far higher quality than is necessary on most consultations.

Similarly, most numbers in use, for policy-purposes, or even in science, are applied in a loose and non-critical way. This is why the whole technological/social system works as well as it does, in spite of the indifferent reliability of most of its information inputs. It has evolved to be resilient and robust against deficiencies of information. However, low quality may manifest in simple error, in a covert rhetorical or symbolical function, or in the ubiquitous hyper- or pseudo-precision of quantitative information. Bad craftmanship in the sphere of information is analogous (and related) to that in production, distribution and administration; and although our institutions and societies generally muddle through, for better or worse, sometimes bureaucratic fantasies can no longer be sustained, and reality crashes in.

Criteria of quality can sometimes be quite demanding, for maps as for numbers. This is most easily appreciated with large-scale maps as used for planning; an error on the ground of a few meters can lead to a misaligned motorway, or expensive compensation for lost property. Similarly standards used in calibration, particularly for high-precision instruments and machinery, determine the dimensions of products in a widely ramified series of subsequent operations. Thus the criteria of adequacy are enormously variable, depending on particular sorts of use. To assume in advance that a low reliability will suffice for all possible areas, is to endanger some unanticipated user. The process of quality-assurance is at the centre of our technological systems, as we see repeatedly when disasters occur. They are equally central to the technology of the information on which the industrial system depends.

6.5. INTERMEDIATE CASES: THEME AND GRAPH MAPS

Hitherto we have discussed the two extreme cases, of maps and numbers, showing how there are strong analogies as well as suggestive contrasts, between them in relation to their characteristic uncertainties. Now we proceed to analyze a few intermediate cases; strengthening the argument that numbers, like maps, are products of design, where quality depends partly on function and on the management of uncertainty. The concept of a map has evolved to include those highly stylized representations, where topographical verisimilitude is nearly completely sacrificed in the interest of a vivid, almost universal and easily grasped picture of some special feature. The archetype for this is the London Transport Underground map (Fig. 3) where the Bauhaus aesthetic achieved a revolution in graphical communication, making it an early classic of contemporary popular design, along with the Coca-Cola bottle and the later VW "Beetle" car. Noone now complains that, in the interest of a clear layout for the Circle and Central lines, several pairs of underground stations, neighbours on the ground, are portrayed as remote (as for example, Bayswater and Lancaster Gate, and Monument and Bank). It has even been suggested that for people who rely on the Underground system (particularly tourists), their effective locational reality becomes that of the Underground map. For a city whose centre has so little overall plan, the design exercise of a graph-map required extreme solutions. Elsewhere, design compromises have been easier; thus in classically planned Washington D.C., the "Metro" system map can be superimposed on the basic topography without noticeable distortions of grid. Even there, the designer's drive for a simple aesthetic produced a slight straightening out of some lines on the map. There

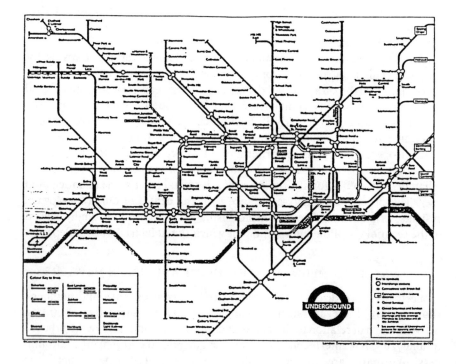

Fig. 3. London transport underground map.

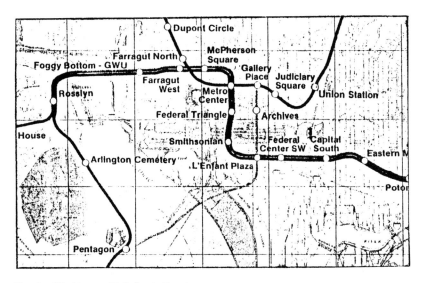

Fig. 4. Washington D.C. "metro" system map.

is no doubt that the later version, where the Blue line does not swerve at all (its exact track being contained on the map within the highly inexact broad path) has as much relevant information for the traveller as the earlier, precise version (Fig. 4 and 5). As we go to press, we can report a new design compromise on the London Undergroung map: the Central line has lost its classic straight path, and at Bank now dips two-thirds of the way down to Monument so that the connecting footway is of a plausible length.

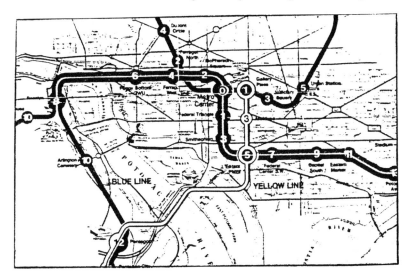

Fig. 5. Washington D.C. "metro" system map (revised).

With the new freedom of map-making from classic representational realism, many sorts of experiments in pictorial representations of data are possible. Where quantitative data are organized by theme, there can be a map whose units have the dimensions proportional to the size of their entry (Kidron and Segal, 1981). Thus on such a theme-map, if it is organized by population, China will be about four times as large as the Soviet Union or the United States. The information as presented by such a map is of very low resolution. First, there is an inevitable perceptual ambiguity involving size: is the quantity proportional to the linear dimension, or to the area? Then, because of the odd shapes and locations, there can be no precision at all in the estimation of sizes. However, we cannot say that the information is of low quality. We can imagine theme-maps being used first to draw attention to important or surprising discrepancies between units in relation to the statistics. For a first survey of a scene, or even for a reminder that some particular country or region is not totally negligible in some respect, theme-maps are genuinely useful. Provided that the numerical information supporting the representation is available on or near the map, their inevitable ambiguities and distortions are not seriously misleading. Thus, even though theme-maps depict information in ways that are imprecise and also unreliable for some uses, still their quality as information sources can be fully acceptable.

Both these intermediate examples show how a discarding of information of one sort (in this instance, topographical verisimilitude) is a valid part of a design exercise where some special function is to be satisfied. In these cases, size and shape are sacrificed, in the interest of either clean design or portrayal of quantitative information. They are particularly useful for our argument, since they show clearly, needing no special expertise for its appreciation, how "form follows function" in the design process. A similar point applies to numbers, and (perhaps surprisingly) to scientific theories. In the former case, we are familiar with the use of binary digits for computers, and their intrusion on human users when a translation is uneconomic. For the human, they are of low quality for representation and calculation. Long strings of 0's and 1's are (without special training) difficult to interpret at a glance, and perceptually confusing: thus, for example, 101 has no immediate significance on its own; it must be laboriously translated to 9 in order to be graspable. Such considerations of design quality are not relevant to computers: they operate by an on-off technology, whose appropriate formalism has a binary logical structure. The psychological operations of "grasping" and "interpreting" do not occur.

A rather less obvious analogy arising from maps of this intermediate sort, is that of scientific knowledge. In a classic exposition of the philosophy of science, theories were compared with maps (see Toulmin, 1953). But these were simply seen as networks on an empty space, and their characteristic distortions and their complex boundaries with ignorance were not mentioned. The process of science was envisaged as a simple increase in the density of information on the sheet; the ideal, in such an analogy, could be the "one mile to the mile map" imagined by Lewis Carroll. Recognizing maps as products of design, we can modify that analogy, so as to enhance our appreciation of the specialized character of scientific models. We refer to these rather than to "theories", to emphasize the more general character of our concern, including conceptual objects and data, as well as formal statements. When we compare scientific models of the same phenomena in an historic sequence, sometimes we are reminded of the steady infilling process of a series of topographical maps, but sometimes also of a collection of variously distorted representations in theme-maps. The same point holds even more strongly when we consider alternative models utilized for studying a complex problem (as in environmental pollution) at different levels or in different aspects. Here the theme-map analogy is quite illuminating. What are labelled as "the same" things are strictly "incongruous"; they cannot be simply superimposed, in terms of data, instruments, methods, theories and results. The different disciplinary structures are, in this analogy, represented by the different sizes and shapes of the "same" thing on the different theme-maps.

6.6. GRAPHS

We can now move another step away from standard topographical maps, towards numbers. We consider graphs of quantitative information, where any representational elements are used purely for visual effect, rather than for conveying information. For our present purposes, the most important feature of these graphs is their inexactness. In a previous chapter we reproduced the graph of successive recommended values of the fine-structure constant α^{-1}. How many readers considered it necessary to apply a ruler to the graph in order to translate the visual information into numerical form? Doubtless very few; the extremely general, coarsely quantitative information conveyed by the graph, was sufficient for our argument.

By contrast, if we had provided that information in tabulated numerical form, we might have retained two or even three digits for successive values and their error bars. From this we may conclude that much tabulated numerical information has more precision than necessary for many of its functions. For purposes of re-calibration of the measuring instruments for α^{-1}, all the information that is available will be relevant; but for illustrating the surprising way in which physical constants can jump, the very coarse topology implied in the graph is adequate, and also appropriate. After all, the real purpose of the graph is to show the effects of systematic error, and it does this by the juxtaposition of error bars with the jumps in the recommended values.

Graphs belong to the same family as maps, in their way of conveying information; much is done by context and implicit suggestion. This extends to the topology, or grid, which defines the inexactness of representation. For numbers, this is done explicitly by the last decimal place (together with various supplementary devices, such as the $\pm n$ conventions). For graphs, the precise value of an element, together with its inexactness, will be given explicitly only in those instances where a fine-mesh grid is superimposed; and it can then be so obtrusive as to interfere with the visual display. The inexactness in graphical information results only partly from that of the original data; that is likely to be swamped by the vaguely defined inexactness of the implicitly suggested topology of the graph.

The suggestion of the topology is one of the elements of graphical design, combining techniques and easthetics, which are so well described in Tufte's book on visual displays of quantitative data. Taking the axes of the graph of successive recommended values of α^{-1} as an example, we see a few hints, which are adequate for the degree of inexactness that is appropriate for the message of the graph. There the scale is given in 10's of ppm's, while the quantitative information is given to seven digits, that is 1 ppm. By interpolation, we can estimate a value perhaps to the nearest fifth of a scale-division; but to estimate the length of an error bar or the size of a jump by naked-eye judgement is highly inexact. The *comparison*, however, is straightforward, since the relevant pairs of error-bars and jumps are juxtaposed. This is what the graph is all about; and so the coarse topology suffices.

Graphs may be used to protect information from hyper-precision of expression. In a previous chapter we analyzed the representation of numerical data and aggregation procedures of the Hazardous Wastes Inspectorate (UK), as described in their report. We commented on the hyper-precision of various entries and the pseudo-precision of the aggregated total. Had the table been given in graphical form, the entries on Table I would have appeared as displayed in Figure 6. The grid that is appropriate here is some multiple of K-tons, 10 at least. Then we see that Cornwall is invisible, Wiltshire negligible, and Cleveland about half Cheshire. For the GLC, we may also represent the extreme unreliability of its data by a bar with the weakest shading, as a reminder that it is the most unreliable. In general, the graph conveys the coarsely quantitative comparisons between WDA's which are appropriate, together with the crucial qualitative information about unreliability. Who would consider this as a less scientific form of representation than the display of the digits, as in the original table? Would any competent decision-maker

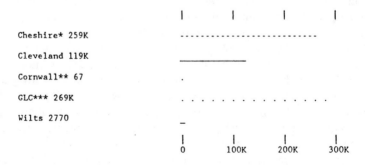

Fig. 6.   (See Table I, p. 51).

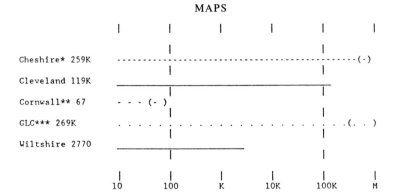

Fig. 7. (See Table I, p. 51).

prefer the hyper-precise numbers to the suggestive graph? We can reinforce this argument, by considering the usefulness of an alternative graphical representation.

Figure 6 was designed to show the relative contributions of the various units to the overall total. In this, the smaller units were visually swamped by the large, even by their uncertainties; this is a result of the choice of a linear scale for representation. But the user of the information may have a different problem, for which they require to compare the returns for all units, large as well as small. In this new instance, a graph on a logarithmic scale may be more appropriate. There is always a danger of an inexpert user being deceived by this representation, in particular, by the apparent similarity in length among all the quantities. The design of graph then must balance these aspects of appropriateness and possible confusion. An emphasis on the exponentially increasing scale on the horizontal axis, will help the inexpert user (Fig.7).

6.7. DIALS V. DIGITS

All the qualities of graphical representations are well illustrated by dials. In one sense, they are merely linear scales bent into a circular form for convenience; and so they appear on a great variety of measurement instruments and gauges. But they show to best advantage in the representation of periodic phenomena, particularly time in ordinary use, as displayed on clocks and wrist-watches. This graphic design presupposes users who are skilled in the interpretation of a minimal set of conventions: the two hands of unequal lengths, and the "12" with its many meanings, at the top. This skill has become so completely internalized in advanced societies, that airplane pilots (as popularized by Hollywood war films) adopted the "o'clock" convention for relative direction, and this is universally and unselfconsciously understood. This case shows a full circle of interpretation: from the graphic display of number (as, a 60° *clockwise* displacement standing for "2 o'clock"), to the numerical description of a visual directional estimate (as "2 o'clock" standing for objects located at 60° to the right of straight-ahead).

On timepiece dials, information is conveyed with the utmost economy of form. The patterns, including direction (based on a "top"" orientation) and several angles (those of the hands, separately and in comparison) enable a fairly precise reading to be made with minimal concentration, even using the smallest set of fixed guides. Thus some clocks dispense with all markers except the one indicating the vertical direction. Any uncertainty about the direction of the long hand, is resolved by the related direction of the short hand.

With the development of the domestic electronic industries, and the displacement of European mechanical gear-driven timepieces by mainly Japanese electronic devices, changes in fashion seemed for a while to render dials in clocks dispensable. The hi-tech aesthetic favoured displays reflecting the enhanced accuracy of the mechanism, even in the cheapest of models. Also, with the possibility of nearly unlimited precision in readouts, users achieved a sense of participation in something excitingly scientific. Even ordinary times of day were commonly read out with a precision more appropriate for describing completion-times of races. People who had been content that their old watch "kept time" sufficiently well so that they did not miss trains, now obsessively checked their digital watch against the standard time-signal, and become anxious if it lagged by as much as twelve seconds.

All this sudden concern for high-precision and accuracy had no relation to the unchanged everyday functions of watches; but was generated by the change in technology (that of the first "LCD" displays developed for readouts of information from microchips). As part of a general trend to digital rather than graphical representations of information, it brought with it a change in the intuition of time: from the continuous, dynamic and contextual, to the discrete, static and atomistic. This created some practical disadvantages. A watch dial has so much extra information given implicitly, particularly about the *region* in which the quantity lies, that the merest glance suffices for a reading. With digits, by contrast, full concentration on all the digits, themselves not endowed with high contrast against the background on the readout panel, is necessary for *any* reliable information about time. This need for concentration, indeed reflection, exposes the pseudo-precision of the information conveyed by the digital watch. For by the time the interpretation is complete, some seconds (and also some hundreds of hundredths) have elapsed. Of course, high-precision measurement of time is frequently necessary; but we notice that in usual practice, this is of a time-interval, with a start (as the beginning of a 100-meter sprint), and an end. Such measurements are subject to all the standard uncertainties of measurements that we have already discussed; but at least their precision is not merely apparent. More recently, fashions have changed back. The style which had always ruled for the very highest and most exclusive class of consumer goods, namely holding to the "natural" (as with cotton and wood over synthetics), re-asserted itself on the mass markets (especially when the cheap digital watch market was saturated). The mechanisms were adapted to readouts in dial form. Now watches can have *both* forms, along with all that indispensable extra information which only microchips can provide.

This little adventure in numerical fashions also illustrates a philosophical point about measurement: the dialectical relation between the discrete and the continuous. Time is, after all, the paradigm case of continuity, captured by the image of "flow". Yet its management has always required the introduction of discrete elements, conceptual for the counting of units, and mechanical for accuracy of measurement. The first working device for converting continuous forces into discrete and very regular motions was the "escapement" for pendulum clockes. Even though the (optional) "seconds" hand in traditional watches generally moved by discrete jumps (as well as the "minutes" hands on very large clocks), still it has required digitalization to bring out so clearly the discrete character of time-measurement.

Dial and digital displays can be seen as complementary. Even on a radio, most tuning is conveniently accomplished on the dial, perhaps as assisted by a signal of quality of reception. But some FM Stereo transmissions have an extremely sharp band-width; the inevitable inexactness of a manually operated tuning knob is too great, unless the scale were to be made inconveniently long. So a digital readout of frequency quite compactly expressing parts per ten thousand, coupled to a digital tuner, is most appropriate here. This complementarity of graphical and numerical representations can be seen in many examples; that of timepieces is particularly useful because of its general familiarity and widespread application. No form is "better" in any absolute sense; the choice of one form, or combination of forms, is a design exercise. Its elements will depict the relevant aspects of the underlying reality, as conditioned by the variety of functions, and anticipated conditions of use, of the device.

6.8. THE FRUITFUL VAGUENESS OF MAPS

The particular excellence of graphs is to enable users to grasp the presence of patterns; these may be derived from a comparison among a few elements, or they may indicate a progressive trend. Numbers are not well suited for this; only when some numbers in a table have many more digits than the others, is there a crude sort of histogram effect. It requires a developed skill to translate digits to magnitudes in such a way that patterns emerge; hence the use of "incomplete-sequence" problems in intelligence tests. The information in graphs depends less on a set of absolute magnitudes than on the patterns formed by their pictorial representations. Patterns depend for their recognition not on precise boundaries, but on relations among structural elements that may individually vary quite widely. As graphs convey information mainly through patterns, that information may also be vague without any loss of definiteness in some particular message; we saw this in connection with the graph of the successive values of α^{-1}. The vagueness of graphs can be fruitful, analogously to the fruitful vagueness of ordinary language. Thanks to its patterns and vagueness, the graph suggests rather than proves, and does so with great economy of elements and flexibility of form.

In contrast, numbers are incapable of being vague, however inexact the information they are used to convey. Even the numerical expression for "error", as for example $\pm .05$, which refers to vaguely defined magnitudes, has a precise form. On the other hand, graphs which convey vagueness will have an inherent limit to the precision of their information. Here we see a polarity of the forms, numbers and graphs, corresponding to the complementarity of the attributes, precision and vagueness. This complementarity goes further still. We have seen that since precision cannot be absolute; it is always characterized by an interval of inexactness, whose endpoints are necessarily vague. Conversely, when a boundary is vague, there must be some limits on that vagueness, lest the opposites melt into each other; and these limits, however arbitrary their location, must be definite, and so must be given in a relatively precise form. In some compositions, the vagueness is so extensive that obliterates any natural boundaries, then we have the powerful perceptual confusions of the topological paradoxes of the kind seen in the Mobius strip (where up-down is lost), the Klein bottle (inside-outside) and the M.C. Escher prints (figure-ground) (see, for example, Escher, 1972). In those cases the perception is drawn along some continuous path, until suddenly viewers find themselves in a radically different situation. In each case there is a transition from a well-defined state, through an intermediate state which is both vague and ambiguous, to the other polar-opposite state, through a vague boundary zone. Paradoxes as these remind us that reliable information is conveyed in ways that necessarily include these complementary aspects; what is appropriate in any given case, providing the best quality for a given function, will be a matter of design, harmonizing the precise and the vague elements in any given scheme of representation.

Neither maps nor numbers, nor indeed any of the intermediate forms, have been designed specifically to convey the various sorts of uncertainty. As we have seen, maps incorporate uncertainties naturally into the information they present, but at the cost of a loss of precision of statement on any quantitative detail. On the other hand, numbers are quite precise in details, unavoidably so; and their relation with vagueness and ambiguity is complex. To design a notational scheme which incorporates both precision and vagueness in a controlled way, and exhibits the various sorts of uncertainty, is the design exercise we undertook with NUSAP. What we have explicitly carried over from the graphic side is minimal: a place-value system for its different categories. But the freedom in expression offered by the variety of notations within the NUSAP scheme enables it, even though cast in numerical or symbolic form, to have a flexibility approaching that of maps. NUSAP as a scheme is the result of a design exercise with graphic elements firmly in view; and every user who constructs their own notation will be engaging in an individual creative activity.

CHAPTER 7

MATHEMATICAL NOTATIONS: FUNCTIONS AND DESIGN

It is not difficult to appreciate how design is involved in graphic productions, as with maps of all sorts and with visual displays of numerical information. Design is less apparent in the case of mathematical symbols, except perhaps in the choice of typefaces in printing. However design is there, primarily involving intellectual criteria but with a significant aesthetic component as well. In this case, the "marks on paper" are intended to be copied, re-copied and combined freely with others, as in a mathematical argument or calculation. Each type of mark, or symbol, represents a concept; in this way it abbreviates the expression from its prose form, encapsulates its meaning, facilitates its use, and interacts fruitfully with the concept it represents.

In graphic productions, each exercise produces a novel product, different in some respect from all that have gone before. In the case of mathematics, where the stock of existing symbols is limited and not enlarged at will, "design" usually means a choice among existing possibilities. It is still justified to speak of a "design exercise", for the task is a matching of forms to functions, in which some aspects are optimized at the expense of others. In the past, such operations in mathematics would not have been conceptualized as design in this modern sense; but we can analyze them in such terms, as implicitly and unselfconsciously doing design. But this means we can obtain clues to their success and survival. Of course, there will occasionally be attempts at creating new symbols or modifying existing ones, in which design considerations are explicit. Our own work is of this sort; and we invoke the idea of design partly to explain what we are doing in the NUSAP notational scheme.

7.1. MATHEMATICS AND SYMBOLISM

In an important respect, mathematics as we know it today is coextensive with symbolic argument and calculation. To be sure, there is always some prose involved in every mathematical text; but even this prose may be abbreviated, using special symbols. Nevertheless, the heart of most mathematical discourse is the manipulation of symbolic language, as a "calculus". There the rules of combination and transformation are fixed, and the meanings of the symbols are generally relevant only at the beginning and end of the formal derivation. It is possible to trace important phases of development in mathematics through a succession of symbolic forms, which steadily replace verbal formulations and becomes more like a true calculus. This process is most impressive in the case of algebra in the sixteenth and seventeenth century, culminating in Descartes' notation ($x,y,z,...$ for variables and $a,b,c,...$ for constants), substantially that in use today. The following examples, given by L. Hogben (Hogben, 1968, 259), illustrate the transition from pure rhetorical algebra to modern algebraic symbolism:

> Regiomontanus, AD 1464,
> > *3 Census et 6 demptis 5 rebus aequatur zero*
> Pacioli, AD 1494,
> > *3 Census p 6 de 5 rebus ae 0*
> Vieta, AD 1591,
> > *3 in A quad – 5 in A plano + 6 aequatur 0*
> Stevin, AD 1585,
> > $3 \,②\, - 5 \,①\, + 6 \,⊙\, = 0$
> Descartes, AD 1637,
> > $3x^2 - 5x + 6 = 0$

We may notice how the second example is mainly an abbreviation of the first; with the third (a century later) we find the concepts "square" and "plane" (for line), and also an ordering of terms by dimension. The next-to-last examples shows a path to formalization which was *not* followed. Stevin's adherence to an arithmetical concept of quantity (breaking with the geometrical analogy which persisted even in Descartes) led him to a notation of exponents, where the unknown was exhibited in terms of its powers; thus he assimilated algebra to his new notation for arithmetic. Seen as a design exercise, this optimized conceptual coherence and unity, but at the price of intuitive clarity and manipulative convenience.

The conclusion of the sequence of forms with Descartes' symbolism might convey the impression that the goal for all mathematical notations is to evolve into a complete self-contained formalism. Such was the dream of those who, in the early twentieth century, supported the abstract axiomatic approach. They wanted to banish meaning, with its associations with non-logical and contradictory intuitions, from the essence of mathematics. Thus, geometry was translated into a fully abstract symbolic system, so as to avoid any reference to space as perceived by ourselves. Thereby, the supposedly paradoxical theorems of non-Euclidean geometry could be rendered innocuous to mathematics. This movement changed the meaning of geometry, separating its completely from "earth-measurement"; it also affected the Kantian philosophy as it concerns the role of Euclidean geometry in our process of knowing (Kline, 1980, 69–99). As a result, the gates were opened to a host of new developments in the philosophy of mathematics, including the problems of the nature of mathematical truth, and the relation of mathematics to the empirical world. This latter issue was part of the background to Einstein's scientific achievement in relatively theory. Thus, symbolism is far from being a matter of mere convenience and abbreviation. The interaction of symbols and concepts is a driving force in mathematics, as well as in the fields of application of, and reflection on, mathematics itself. The apparently austere and abstract world of pure mathematics has a creative dialectic between symbol and concept, just as creative human thought of any sort.

The process of symbolization in mathematics is not uniform; nor indeed should be so. As mathematics is realized in practice, there are many contexts of application, each with their own relevant users and meanings, all coexisting and interacting, so that any particular symbol is the current outcome of a complex historical process. To fix a particular symbol rigidly to a concept, may well reduce ambiguity and vagueness; but this will incur severe costs in flexibility, fruitfulness and power of communication.

7.2. DESIGNING FOR UNCERTAINTY

Our present task is to design a notational system with the new function of incorporating the different sorts of uncertainty, which are so important in policy-related research. As we have seen, uncertainty has been treated mathematically from the seventeenth century onwards. First through probability theory, and then statistics, different aspects of uncertainty have been expressed in symbolic form and included in calculi. This has been accomplished in spite of the continuing well-known difficulties in the definition of basic probabilistic concepts (such as "random") or in the choice between different conceptions of "probability" itself (see Lucas, 1970, 213–215). In this aspect, the treatment of uncertainty has followed that of "the calculus", where an elaborated and powerful development could proceed (as in the eighteenth century) in the absence of clarity on foundations; one of the main foci of controversy then was the concept of "differential", expressed in Leibniz's symbolism as dx. Most recently, some aspects of uncertainty relevant to policy-related research have also been mathematized (as in Bayesian statistics or fuzzy sets theory). These formalism suffer from two characteristic difficulties. First, they fail to distinguish among different sorts of uncertainty; they merely represent judgments of any sort of likelihood, or measures of any sort of vagueness, respectively. As a result, they are not well adapted in themselves to express crucial distinctions. In consequence they are vulnerable to the paradox of infinite regress, as we have seen. This paradox does not represent a merely philosophical puzzle; it demonstrates their inability to provide guidance to users on how to interpret and when to over-ride conclusions derived from the formalisms. The second difficulty is that the residual uncertainty remaining in both kinds of formalisms, must be expressed in prose; this is the only way to avoid an infinite regress. Thus, in terms of the programmes in which they were conceived, these symbolisms are self-defeating. They do not provide a self-contained algorithm for the management of uncertainty.

One of the most important functions of the NUSAP notational scheme is to overcome the dichotomy between the hard quantitative information given formally, and the soft qualitative information that is expressed in prose. Only when it becomes standard good practice for all quantitative statements to include uncertainty, will scientific information be well managed in this respect. To this end NUSAP distributes the three sorts of uncertainty (inexactness, unreliability and border with ignorance) among its latter categories. In this way, unlike when a limited symbolism coexists with prose, the absence of any entry is automatically signalled by an empty space in a string. In this respect, NUSAP operates like the blank spaces in the old maps: it makes us aware of our ignorance. Otherwise, we are at risk of believing that we know more than we do, a dangerous state of affairs.

As a product of conscious design, NUSAP has given greater priority to transparency and convenience of representation, and to the integrated display of all the sorts of uncertainty, at the expense of a facility for simultaneous calculation with all the sorts of uncertainty (for which no existing notation is adequate). We never lose sight of our ultimate objective for NUSAP: that it should facilitate communication of quantitative information (with a special focus on uncertainties), and thereby help to maintain and enhance quality assurance in this very broad area. The blend of symbolism and informality in the NUSAP notational scheme will also contribute to the development of the skills necessary for the process of quality assurance. We have already explained some of the principles underlying the design of NUSAP. We shall later articulate the criteria on which its design has been based; and show that these are entirely natural. Most of them are implicit in the design of notations for numerical systems, and the others are related to the special functions of notational systems designed for the practical management of uncertainty.

7.3. FUNCTIONS OF NUMBERS

In order to establish the relevance of design to mathematics, we shall first review the various functions of numbers, and show how the choice of notations in special contexts reflects design considerations. Some of these functions are located in the prehistory of numbers themselves, but we are not going to discuss them in a speculative chronological order. First, we have the primitive *distinction of multiplicity*. At its most basic, this is 1–2; and then, 1–2–Many. This distinction was, at the beginning of this century, the basis for the "intuitionistic" school of foundations of mathematics; "twoness" is the elementary iterative property for the generation of numbers, whereas "oneness" is not. Then, we have 3; and it is interesting to note that the words naming 3 in European languages (three, drei, tres, tre, trois,...) have the same root as the Latin trans, meaning beyond, denoting the jump from 2 to Many.

Before there were extended names or proper notations for representing numbers, there could be systems for *tallying* collections, as by notches on stiches or knots in strings. These could themselves become quite sophisticated, as in the Inca civilization of Peru, in South America (Cajori, 1928, 38–40). For the achievement of proper numerical systems, two complementary functions are necessary: *counting* and *naming*. Both involve an abstraction, from the properties of particular collections to the grasping of a general property of multiplicity as quantified. It implies the transformation of the number-words from adjectives to nouns in ordinary languages. This process of transformation need not go to completion, as some modern languages retain separate number-words for different sort of things; Japanese is a case in point (Wilder, 1968). Also, the use of prose rather than numerical symbols for the names of numbers is necessary for ordinary discourse; and since such sorts of names do not always possess iterative properties, interesting ambiguities arise, as in the case of "billion" for 10^9 *or* 10^{12}. Closely allied with the naming function, is that of *magic*; this survives today in superstition as with the number 13; or in religion as with the numbers 3 or 7 (Davis and Hersh, 1981, 96–108).

We can now discuss some of the remaining functions of numbers, proceeding from the less formal and quantified, to the more so. First, we have a general function, we call *indexing*, which can be analyzed into three sorts. The least structured is *individualizing*, as when numbers are used in an arbitrary way, as for marking the shirts of members of a sports team (there may of course be exceptions, as particular roles may have special numbers; but these are more related to tradition than necessity, as for example, the number 1 for the goalkeeper in a football team). Next is *locating*, as with rooms in a large building or hotel. The number helps the user to identify, for example, the floor and relative position, but room 306 is not necessarily smaller than or inferior to room 307. Such locating indexes are very common; we use them in numerical form in library classification systems, and in alphabetical form in directories and encyclopedias. The locating index does impose an order on its elements, but it is arbitrary or purely conventional in that it does not represent any inherent quality, and it can usually be replaced by others for the same function. A stronger sort of indexing is *gauging*, as when the ordering represents an attribute that is seen to relate more essentially to the elements in the collection. Numbers may well be used here, though without proportionality or measurability. Thus, we have the number in the Mohr's scale of relative hardness (substance A is harder than substance B, if A scratches B, but B does not scratch A). In this case, "hardness" is perceived as a more essential attribute of a substance than the initial letter of the name of an encyclopedia entry in a particular language. A more sophisticated example of a gauge is the Apgar Index, an indicator for a baby's condition at birth, where five attributes (colour, muscle-tone, response to stimulation, respiratory effort and heart-rate) are scored visually as 0, 1 or 2, and then added to provide an immediate assessment of the general health and viability of the baby (Thomson, 1979). Not all simple indicators are indexes in this weak sense; thus the Richter scale for earthquakes is actually a logarithmic measure of the energy released, and so is a synthesis of refined physical measurements.

These three sorts of indexing impose different criteria for the choice of their notations. For individualizing, there may be an intention to *avoid* suggestion of inequality (as among groups of children in a school); then the indexers may need considerable ingenuity to find sets of objects (colours or geometrical shapes) where no ordering can be inferred. With location, an indexing set which avoids the implication of being an arithmetic may be desired; thus a mixture of letters with numbers may be adopted. For gauging, on the other hand, metric notations are important, and so numbers would be used; here conventions and rules may be needed to prevent hyper-precision, as the calculations can easily produce more precision than is justified by the underlying quasi-quantitative data.

Next is *estimation*, which might be considered as measuring without actually performing the physical operations of measurement. This may appear in practice as the "back-of-envelope" calculations, with which certain experts (as in engineering and architecture) do preliminary rough and ready calculations, providing users with an order-of-magnitude evaluation. Bayesian statistics can be seen as a sophisticated development based on the estimation of personal probabilities. (These estimates are not merely of a single quantity, but take the form of probability distributions, elicited under expert guidance and following explicit rules in order to avoid contradictory results.) Uncertainty requires skilled management for its estimation to be effective. Simple one-digit calculations, giving an order-of-magnitude result, may be more appropriate in certain contexts than a detailed computation with seemingly precise digits, in which the uncertainties remain masked. Thus, unlike in ordinary arithmetic, the skill of the expert may be very important, not merely for the speed of the work but also for the quality of the result.

In his illuminating paper, F. Mosteller describes a number of techniques for estimation. In many practical and industrial contexts "rules of thumb" are invoked to obtain more appropriate estimates; thus the standard costs per unit area of building are modified by a factor less than one, for attics and cellars. A more complicated example is the estimation of the number of miles driven annually by American automobiles. This may be done on the basis of various plausible quantitative assumptions: from the total registrations and average use; or from the total fuel consumption and average mileage; or from the total of road space. The first two estimates agree to within 30 %, indicating a fairly robust method (Mosteller, 1977). Finally we may remark that order-of-magnitude exercises can provide a challenge to the imagination, not least in representation skills. Some say that the test for identifying potential physicists is their handling of the following problem: How many piano tuners are employed in your city? The calculation may proceed from the estimated number of households and concert-halls, the density of pianos and the frequency of tuning (both very dependent on social and cultural aspirations) and finally the workload of piano-tuners. An important additional skill for estimators is their ability to choose an appropriate notation, so that the uncertainties inherent in the operations can be communicated properly. Ordinary arithmetic is the obvious tool for calculation here, but it must be handled with great care lest it be misleadingly precise.

We come now to *measuring* and *counting*. From antiquity up to modern times, these were accepted as the two main streams of mathematics, geometry and arithmetic, dealing with continuous magnitude and discrete quantity, respectively. In modern times the distinction has broken down in many ways; and since classical geometry is no longer a well-known subject, it is anyway not so meaningful as previously. It does, however, serve to remind us that there is a dialectic of similarity and difference between the two sorts of operations. As to counting, philosophers have explained it in terms of a one-to-one comparison of a set of integers with a set of things "out there". When we run out of things to tally off by the integers, we say that the count is complete; and the last number to be used, defines the size of the set. Such a philosophical account helps to explain the effectiveness of an early system of notation for numerals: the alphabetical. For the letters provide a familiar, ready-made ordered sequence of symbols. Problems of larger aggregated units can be managed with spare symbols and extra marks, like '. Such a notation has the incidental advantage of relating counting to magic, through the "calculation" of names, or gematria. The costs of the design choice include a lack of transparency of alphabetical symbolism for numbers, as compared with dots, strokes, or their aggregations; and (more important for their survival into the future) their inconvenience in general calculation.

When we measure, in the simplest cases we actually do a count. This is of the largest number of the set of scale-units which does not exceed the dimension of the thing being measured. Beyond that it is normal to estimate an extra bit, the size of the remainder of the thing that is not included in the given aggregate of scale-units. This estimation is the interpolation between finest gradations on the scale on which the thing is being measured. Thus measurement involves both the precise operation of counting and the imprecise operation of estimating; it combines both and in a sense lies between. We have already seen that estimation is itself a form of measurement; it happens that most counting in practice is also. For "pure" counting can be accomplished by unaided human work only for very small collections. Others involve aggregated units; and the operations on them are similar to measurements, or even to estimation. Even when the units are indivisible, as heads of population, the gross statistics (as the total population of a city), are not derived simply, as by lining up all the inhabitants and giving each a number. In nearly all cases of large collection of discrete things, whether the units be of currency or of people, what is reported as if it were a count is actually the result of a process involving aggregation, approximations and also estimates by interpolation and extrapolation.

Next we come to *calculating*, which can be seen as a development from the purest form of counting. The operation of calculation came historically even before numerals; in many cultures there were elaborated systems for solving complex problems in area or weight measurement, or even astronomy, which involved rearranging patterns of small objects (thus, "calculus" means "pebble" in Greek). Although the formal rules of calculation in any field are fixed, and they can in principle be carried out independently of the meaning of the object, in practical cases it is otherwise. As we saw in the "fossils joke", there is a need for a supplementary, artefactual arithmetic, when orders-of-magnitude are involved. Thus estimation conditions calculation in practice. This influence of the "softest" on the "hardest" is quite crucial in the proper operation of computers, where (as we have discussed) the management of uncertainty, from the conventions for rounding-off to the evaluation of programs, requires highly developed craft skills.

Thus we may discern three sorts of dialectical oppositions in play: the classic one of the discrete and the continuous; the operational one of precision and inexactness; and finally, the practical opposition between objective repeatable operations, and intuitive and unique judgements. The main functions of numbers generally distribute themselves uniformly along the three polarities, with pure counting at the "hard" end and pure estimating at the "soft". Both of these extreme cases are represented in much ordinary practice; and this raises the question of whether the same design of notation design is appropriate for them both.

Our analysis of the functions of numbers has focussed on those which have been familiar in practice for a long time. Indeed, they seemed to exhaust the possibilities of what numbers can do. But with the growth of the new issues where science and policy are thoroughly mixed, numbers, as we have discussed previously, are increasingly used in a context which is essentially political and rhetorical rather than scientific and technical. The function of *communication* was not previously considered to be of great importance. The expert audiences for quantitative information should not have difficulty in interpreting it; while inexpert, lay audiences are of no philosophical significance. As Tufte shows so well, communication in graphical form is a demanding art, whose neglect leads to error and confusion. Numerical information, when presented in a policy context, may also be fraught with the same dangers; we have shown how skills are required in the provision of information on policy-related research. It is not merely that inexpert audiences may lack the technical background for understanding special notations and conventions. The information as purveyed does more than merely to inform. Numbers can appear in statements of suggestion, advice and command, as well as conveying the assertion of authority and reassurance. Some such functions may be in conflict with those of simple informing; from the cognitive point of view, they may be considered as misleading or misinforming. Indeed, the information skills most relevant for the general public may well be more in the area of quality-evaluation than technical comprehension. This is why traditional notations now need enrichment, to meet these new needs.

It is in connection with calculation, that the importance of design of numbers is most obvious. Digital computers have an off-on logic, and so the representation of numbers on which they operate is binary, using only the digits 0 and 1. For human calculation, on the other hand, such a notation is impossibly cumbersome; its logical simplicity has no benefits commensurate with its costs in loss of convenience and of correspondence with the elementary counting system of our digits. Further it has no relation to our common prose names for numbers, which are decimal. Yet for the calculations done inside the machine, binary is the perfect design. There are compromise numerical systems, such as octal (eight digits: 0,..., 7) and hexadecimal (sixteen digits: 0,..., 9, A,..., F), which translate directly into binary, being as it were "packed" binary systems, and which at least are not so cumbersome, although still somewhat counterintuitive.

The partial success of these particular number-bases contrasts with the perennial failure of the duodecimal (base-12) system, which had been advocated on the grounds of its convenience for manual calculation, particularly fractions, as well as its connection with the measurement of time and angles. For the duodecimals, inherited from the Babylonians, had only a special niche; they could not be extended to general use at the expense of decimals. In the case of the binary-based scales there is no crusade for consistency, only a practical mixture of different bases associated to different functions. On the other hand, the ancient sexagesimal system for periodic phenomena has proved remarkably resilient. The decimal-based "metric" system, introduced during the French Revolution, is now nearly universal; but the reform proposed then of making a right-angle of 100 degrees, as well as a week of ten days, and a day of a hundred units, survive only as curiosities. Even though calculations in traditional systems with mixed bases are incoherent (the sexagesimal seconds (of time and angle) being subdivided into hundredths) the advantage of familiarity seem to be overwhelming. Thus pupils need to master an arithmetic including exercises like:

$$
\begin{array}{r}
44'\,32''.36 \\
+\ 27'\,41''.81 \\
\hline
71'\,73'' + 1.17''
\end{array}
$$

$$
\begin{aligned}
&= \quad 71'\,74''.17 \qquad \text{(Decimal)} \\
&= \quad 72'\,14''.17 \qquad \text{(Sexagesimal)} \\
&= 1°\,12'\,14''.17 \quad \text{(Sexagesimal)}
\end{aligned}
$$

Perhaps it is because such calculations are relatively uncommon by hand, that the incoherence of the system is not a serious design defect.

Calculation was, however, fatal for the survival of the Roman numerals. Even when they were the standard system for naming and counting, reckoning systems, including the abacus, were used for calculus. For naming, the Roman numerals do have some advantages, in the clarity of distinction among different aggregated units: M, D, C, L, X and V show quite clearly that they refer to different numbers. Tradition maintains them in use, mainly where a symbolic significance is desired, as for dates of major events. They can also function as a shorter series of indexing numbers running parallel with an ordinary set, as for volumes or sections of a publication.

7.4. NAMES FOR NUMBERS. THE "BILLION" STORY

As we have seen, the naming function of numbers is quite significant in ordinary practice, even though it may not be relevant for most of science and mathematics. Thus in technical contexts, a "million" is simply understood as 10^6; the prose name carries no special advantage. But in ordinary discourse, such prose names for large numbers, expressing the units of aggregated countings, are important for general comprehension. The design criteria for their composition include convenience and vividness of imagery, along with coherence with the names for smaller units. By this last criterion, the best system of names for large numbers, would be a fully recursive one, perhaps on a base of a thousand; we could have, say, "thousand", "thousand-squared", "thousand-cubed", and then (moving on from the older, geometrical names for exponents), "thousand-fourth", "thousand-fifth", and so on, certainly as far as ordinary usages requires. But such a logical system is cumbersome; and so there was created the "million" ("big-thousand") for "thousand-thousand". Even this loses in its ostensive quality, compared with "thousand-squared". At this point real trouble begins.

What should be the basis for naming large units beyond this "million"? We can choose a logical system, where we proceed by "million-second power", "million-third power", and so on, whose natural names are "bi-million" or "billion", "tri-million" or 'trillion", and so on. The British philosopher Locke used these names in 1690. Such a recursive system is coherent, and gets us to very big numbers quickly; but it has the design defects of leaving very large gaps, to be filled by cumbersome locutions. Thus, the name of 10^{11} would be "hundred-thousand-million". To obviate this, an intermediate unit, of 10^9 was created, called "milliard". For consistency (and especially since very large numbers are increasingly in use) we would need a similar unit at 10^{15}. Would it be called "billiard"?

An alternative design was introduced by French arithmeticians, and adopted in the United States; there the recursive unit is a thousand all along, so that a "billion" becomes "thousand-million", "trillion", "thousand-squared-million", and so on. This avoids the large gaps, but it is clearly incoherent. As the first element is the second power of thousand, the number-names are all different from the associated powers. The choice between the two designs is in one sense purely conventional, as neither system of nomenclature has the slightest importance for science, mathematics or philosophy. But, as settled linguistic habits, they become the property of nations, and their use and diffusion reflects relations of power between cultures and economies. This is most apparent in the case of the United Kingdom, where for some decades the transatlantic influence has been strong; and the old British "thousand-million" has, all unnoticed, given way to the American "billion". Now there are two "billions" in circulation, the French-American 10^9 and the old British 10^{12}; future historians may well be misled by the unannounced thousand-fold reduction in the size of the British billion. The confusion might be resolved by going back to the larger unit of "ten-thousand", as in the classical Greek "myriad". This would have the incidental advantage of harmonizing our numerical systems with the use of the Japanese, also based on 10,000, with "Man", "Oku" and "Chou" as its first, second and third powers respectively.

The discussion about the "billion" may seem amusing or even pointless, until we recall that one of the functions of numbers is the communication of scientific information. When large quantities are conveyed the non-numerical names of numbers are usually preferred to the numerical ones (we say "million" and not "ten-to-the-six"), and if these verbal names are ambiguous, as is the case of billion, then the message becomes confused or misleading. The moral of "the billion story", as with the mixed sexagesimal notations, is that when numbers are used, the criteria of quality of any particular design depend on many factors, some psychological and some historical. Clearly, there can be changes in the successful designs, depending on any one of the sorts of factors. But it is important to appreciate the variety of possible options among designs, and the multiplicity of criteria for such choices. In that way, we can understand the systems in use, make an effective critical evaluation among them, and then devise new systems when the existing ones are not appropriate for new uses. This has been the cause of development through history; our work is intended to extend it to the new functions of incorporating all varieties of uncertainty into quantitative discourse.

The problem of numerical names, where tradition and convenience clash with consistency, is not restricted to the verbal usages of non-experts. Even within mathematics itself, there are important cases of anomalies in symbolic forms. Some important mathematical functions are denoted by symbols which are abbreviations of their prose names rather than completely stylized notations. In this sense they have remained at a phase of development analogous to algebra in the sixteenth century. Thus the logarithmic and trigonometric functions are denoted as by $\log x$ and $\sin x$, and so on. When these functions become elements of a simple algorithm, problems of design appear immediately. In the case of inverses, we may invent a new name, but this reinforces its non-symbolic character, as *anti*-log or *arc*-sin. If we wish to develop the algorithmic character, we need a symbol for inverse adjoined to the function-name; this is usually $\log^{-1} x$ or $\sin^{-1} x$. However, this pseudo-exponent may be confused with the real exponent, in the existing notation for powers, as $(\sin x)^2 = \sin^2 x$, or $(\sin x)^{-1} = \sin^{-1} x$. The point is that *this* design problem has no obvious best solution. To remove the problem by creating totally new symbols, as was done in the case of the "elliptic functions" in the nineteenth century, would be interpreted as an "user-hostile" act, and would certainly fail in the academic marketplace of overworked teachers and inexpert students. Just as in the case of the billion, a deliberate radical policy-changes would need to be justified in the relevant societal terms; and in the inherited design of the symbol, with all its drawbacks, survives by its familiarity.

7.5. PLACE-VALUE SCHEME

The success of our modern decimal system of numbers is based on its recursive properties. These make it very convenient for naming, since quantities of any size may be reached by combinations of digits from the small original set. Also, the numbers in this form are nearly perfectly adapted for calculating, not merely in integer arithmetic but also in the great variety of extended systems, including fractions, negative, surds and complex numbers. At the root of all this power is the combination of place-value and zero. Although such a notational scheme may seem to be so overwhelmingly superior as to be almost a "fact", still it is a product of human invention; and moreover, as a design product does not optimize over all possible functions of numbers. We have seen that for the naming of large units in communication, the ambiguous prose names are still preferred to place-value notations such as 10^9. At a deeper level we have seen the ambiguity of zero, which is essential for its function in the place-value system, and yet which introduces confusion in the context of estimation. We may therefore consider our standard decimal notational scheme as a product of design; this helps us to identify certain key features which will be very important for our later discussion.

The place-value system can be analyzed at three levels. The most general, is the "scheme" level. This is the sequence of powers of the base, defining the places (to left and then to right of the "decimal point") in the string of digits. When the system was extended by Stevin to include fractions, he made this potential meaning explicit by drawing a circle around the number indicating the power, but it was soon realized that these were redundant. Thus in the expression 102.66 we know exactly which power is implicitly expressed by the position of the digit in relation to the decimal point. It is interesting to notice that this notational device, which provides such power and flexibility, uses a visual, graphic technique to convey meaning. The scheme considered abstractly, is a string of places, on either side of a dividing point, in which each place is implicitly defined by a positive or negative integer exponent.

The place-value scheme as defined is a very general notational device; it is not restricted to any particular base of numeration. Thus, the binary notation uses the place-value scheme just as does the decimal. The meaning of a digit in a place in the string thus depends on two factors: the relative position, which provides the power of the base, and the particular base chosen. The choice of base thus defines the level we call "notation", within the general scheme. Finally, when we have particular digits in all the places in the string, in the context of a notation as defined, we have an "instance". We can illustrate the three levels: scheme, notation and instance, using two different numerical bases, as follows:

Scheme	2	1	0	2	1	0
Notation	10	10	10	2	2	2
Instance	1	0	1	1	0	1

To obtain the decimal value of both instances, we have to perform the calculation:

$$1 \times 10^2 + 0 \times 10^1 + 1 \times 10^0 = 101$$
$$1 \times 2^2 + 0 \times 2^1 + 1 \times 2^0 = 5$$

The "instances" used in any calculation must all belong to the same "notation"; otherwise complete confusion would result. We shall use these distinctions later on, in discussing the NUSAP notational scheme.

The system of place-value with zero has another advantage, which is not always clearly appreciated in spite of being widely used. This is its ability to convey meanings which are closely related, and yet distinct for some purposes. It may seem surprising that numbers in standard form should be able to convey nuances; but this is so. Different descriptive meanings are suggested by alternative interpretations of the string of zeroes to the right of the non-zero digits. Suppose we have a 2 followed by six 0's (in base 10). If all the digits are considered as counters, then the number expressed is that conveyed by the sum 1,999,999 + 1. But if some zeroes are considered fillers, say three, then the number can be represented as 2,000K; and so on. Such distinctions may be quite important in practice, as we saw with the "fossils joke". Given this ambiguity in any string of zeroes, making the meaning clear becomes a design exercise, involving the choice of an aggregated unit of counting. What is appropriate will depend both on the nature of the counting operation, and the function of the number as expressed, for communication. Thus, if the number refers to money, and relates to an actual count of notes, the filler digits will relate to the denomination of the notes. For example, in thousand-dollar bills, our number could be expressed as $2,000 \times \$1,000$; or of in hundreds, $20,000 \times \$100$. How to represent the filler digits will depend both on the information and the message to be conveyed. Tables of figures in financial accounts may be headed by "$K" or "000's", the former for a readership assumed familiar with scientific language, and the latter for lay persons. The expert users would also be expected to know when K stands for 1,000 and when for $2^{10} = 1024$, some $2^1/2\%$ greater.

7.6. FRUITFUL CONTRADICTION

The problem of naming the large aggregated units is not merely one of prose style; analogous problems occur in relation to the fertile ambiguities of the zero. It is well known that the place-value system could not support a real "calculus" in the absence of a cypher representing the empty place; yet the zero, once introduced, is obviously a "non-standard" object, having its own special arithmetical rules, such as $n \times 0 = 0$, and $n/0$ and $0/0$ declared meaningless. Thus, zero, as an early extension of the "natural numbers" has paradoxical properties, "monsters", which require *ad hoc* rules so that they can be barred or tamed. Other extensions of the number system produce their own monsters. Thus negative numbers break the standard rules of inequality of fractions, namely that

$$a < b \text{ and } c > d \quad \text{implies} \quad a/b < c/d$$

(We see this if we have $a = d = 1$, $b = c = 2$; the relation is then $\frac{1}{2} < 2/1$). However, if we set $a = d = -2$, then by the inequalities as extended to negative numbers, $a < b$ and $c > d$, but the fractions are $-2/2$ and $2/-2$, respectively, both equal to -1. Thus a very basic rule of arithmetic of earlier ages was violated by the negative numbers. The anomaly is no longer serious, as we do not work with ratios, and so the problem can be suppressed.

The paradoxical properties of zero are not exhausted by its arithmetic in place-value notation. As an element in the sequence of positive and negative integers, it occurs as an exponent. Thus a^0, that is "no" powers of a, is by definition 1. Exponents produce a rich crop of monsters; thus $a^{\frac{1}{2}}$ has not one but two values, for a positive. If a is negative, the commutative rules do not apply, thus $[(a)^{\frac{1}{2}}]^2 \neq (a^2)^{\frac{1}{2}}$; try with $a = -1$ for a test. The zero causes yet more difficulty, as when we write $\log 0$, a meaningless expression; even the expression $\log a$, for a negative, caused great puzzlement since it has an infinity of values.

The purpose of all these examples is to show than the basic concepts of our modern arithmetic and analysis are enormously fertile, capable of extensions (with the same symbols being reinterpreted) to very new domains; but that the price of this power is ambiguity and contradiction, which need to be managed. This is done partly by the introduction of *ad hoc* rules and conventions; and also by directing everyone's attention away from the contradictory features, in order to maintain the ages-old illusion of certainty in mathematics. This can be done even by skillful use of ordinary language, thus the standard description of contradictions as "paradoxes", implying merely "surprise". This linguistic technique for lessening the impact of such negative examples ("monsters") is an ancient and honourable strategy in mathematics, starting with the use of "irrational" by the Greeks to name incommensurable magnitudes such as $\sqrt{2}$.

Thus our modern number system as designed, and the rich algebra built on it, display the fruitful contradictions of an ongoing dialectical process. The same can be seen in the calculus, leading on to higher mathematics. There the association of notations with concepts was particularly intimate. Newton conceived the concept of "fluxion", a flowing quantity described by its velocity (in symbolic form, $\dot{x}$). Its inverse was the "fluent", denoted $\grave{x}$. Immediately we see practical problems of typography, and there were insuperable problems of notation for iterated use. By contrast, Leibniz produced his "differential" dx, whose sum is an "integral", denoted $\int$. Both symbols are easy to print singly, and as iterated; further they are very expressive in their meaning. Moreover they were capable of immediate development and extensions, creating "the calculus" as we know it. The price for this design was the obvious conceptual contradictions of the infinitesimal or infinitely small, quasi-zero "differential" that was essential for the Leibnizian sums; while

Newton's "fluxion" could rest on the intuition of motion and velocity. An outside critic, Bishop Berkeley, was quick to point out all these contradictions in his defence of the religious doctrines, which are "above reason", as superior to the mathematicians' arguments, which are merely "beyond reason". For mathematicians then and for centuries later, he was an irritant, but was never conceived as a real threat. (For a new twist to the story of infinitesimals, see Robinson, 1966).

Later, the complementary tendencies of fertility and contradiction combined with explosive force in the case of the actual infinite. Until the late nineteenth century, infinity had been the worst monster in the mathematical zoo. Numerically derived as the limiting quantity obtained from the quotient $1/x$ when x becomes very small, it was indispensible in the development of the calculus. Great effort in the rigourization of analysis was necessary to tame it, as by Cauchy and his successors. Even before that process was complete, Cantor's "transfinite numbers" opened a new Pandora's box of contradictions. These then led to the creation of important new fields in mathematics, and through them to the final destruction of the millenial ideal of certainty in mathematics.

7.7. SYMBOLISM IN CHEMISTRY

This story shows us two related dialectical processes in play in the development of mathematics: that of fruitfulness and contradiction in concepts, and the interaction of concepts and symbols in operations and extensions. Such processes are not limited to mathematics; they occur wherever there are conceptual structures capable of representation in symbolic form. A paradigm example outside mathematics is chemistry, relating to the names of the elements. In alchemy, the symbols themselves were considered as having magical properties. Once a scientific chemistry was established, there were many attempts to design appropriate symbols. Such symbolisms attempted to convey the names or meanings of the elements, and also to represent important relations among them (in the eighteenth century, these were mainly elective affinities). Thus John Dalton's atomic theory was embodied in a "chemical philosophy" complete with graphic symbols and rules for their combination. They had the qualities of transparency to a high degree; but they did not enter widespread use because of typographical difficulties. Through the nineteenth century, philosophical rigour in the graphic design of symbols gave way to practical convenience, in the use of abbreviations for the standardized names of the elements on Lavoisier's system. Although the symbols never become a calculus, they have always been extensively used both for reaction-equations and for depicting structure.

The great scientific breakthrough of graphics in chemistry came with the "periodic table" of Mendeleieff (Figure 8). In it we find a place-value notational system, involving a strictly ordered numerical sequence of elements which winds its way through the columns. The location of any element within the

Period	I	II	III	IV	V	VI	VII	VIII			b(0)
	a · b	a · b	a · b	a · b	a · b	a · b	a · b	a			b(0)
1	H 1.0										He 4.0
2	Li 6.9	Be 9.0	B 10.8	C 12.0	N 14.0	O 16.0	F 19.0				Ne 20.2
3	Na 23.0	Mg 24.3	Al 27.0	Si 28.1	P 31.0	S 32.1	Cl 35.5				Ar 39.9
4	K 39.1	Ca 40.1	Sc 45.0	Ti 47.9	V 50.9	Cr 52.0	Mn 54.9	Fe 55.8	Co 58.9	Ni 58.7	
	Cu 63.5	Zn 65.4	Ga 69.7	Ge 72.6	As 74.9	Se 79.0	Br 79.9				Kr 83.8
5	Rb 85.5	Sr 87.6	Y 88.9	Zr 91.2	Nb 92.9	Mo 95.9	Tc	Ru 101.1	Rh 102.9	Pd 106.4	
	Ag 107.9	Cd 112.4	In 114.8	Sn 118.7	Sb 121.8	Te 127.6	- I 126.9				Xe 131.3
6	Cs 132.9	Ba 137.3	La 138.9	Hf 178.5	Ta 180.9	W 183.9	Re 186.2	Os 190.2	Ir 192.2	Pt 195.1	
	Au 197.0	Hg 200.6	Tl 204.4	Pb 207.2	Bi 209.0	Po	At				Rn
7	Fr	Ra	Ac								

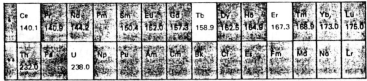

| Ce 140.1 | Pr 140.9 | Nd 144.2 | Pm | Sm 150.4 | Eu 152.0 | Gd 157.3 | Tb 158.9 | Dy 162.5 | Ho 164.9 | Er 167.3 | Tm 168.9 | Yb 173.0 | Lu 175.0 |
| Th 232.0 | Pa | U 238.0 | Np | Pu | Am | Cm | Bk | Cf | Es | Fm | Md | No | Lr |

Fig. 8. The periodic table. Elements shown shaded were unknown at the time the table was constructed.

matrix conveys important information about its properties. The empty spaces on the table (or map) provided clues for the discovery of new elements; for a long time, the periodic table played a crucial role in research by the visual definition of the border with ignorance. The symbolism of chemistry could never be fully formalized, as for the establishment of an universal algorithm. But the symbols as embellished with extra information of many sorts, could function effectively within a simple calculus and simple graphic representations of structure. (For a full account of the earlier symbolic development of chemistry, see Crosland, 1962).

The example of chemistry shows that a place-value notational scheme can be heuristically useful and capable of extended scientific development, even when it is not a fully-fledged calculus. We will show that for the management of uncertainty, the NUSAP notational scheme has analogous properties.

CHAPTER 8

THE NUSAP SCHEME: INTRODUCTION

The new tasks of science, particularly policy-related research, present new requirements for an appropriate mathematical language. Previously, notations were generally intended for use by experts, working in problems defined within specific fields of pure or applied mathematics. The assessment of quality of results, and communication to lay audiences, were considered to be outside the realm of scientific practice, to be accomplished by informal means, and not requiring a special technical language.

Uncertainty has been managed and expressed, as by statistics, on the assumption that it is a purely technical issue, soluble within the framework of mathematical approaches and notations. Thus classical probability theory may be interpreted as a branch of algebra, and statistics as a branch of analysis. Graphical methods in statistics are themselves an application of the principles of co-ordinate geometry. In relation to communication to a broader public, the main task has been seen as conveying the certainties, not the uncertainties, of science. Only when statistics is involved in statecraft, is there believed to be a need to translate between the cognitive and the policy contexts; and in general the sophistication of this information was modest.

Under our new circumstances, the tasks of policy-related research now include the management of uncertainty, the assessment of quality, and communication with lay publics. In addition, enhanced public participation in policy processes requires the diffusion of skills in all these. This must be done in the general cultural context of the decline of a simple faith in Science as necessarily purveying truth, and in scientists as being competent, disinterested and truthful by the very definition of their calling.

An appropriate new mathematical language for science will therefore be concerned with the provision of conceptual and notational tools designed to enhance where possible, and to create where necessary, the skills required for science in the policy context. These skills will include the management of all sorts of uncertainty, so that uses of information, lay and expert alike, can appreciate the complexity of the production process of scientific information. Similarly, skills in the assessment of quality of information as an input to the policy process, will involve an awareness of the interaction of its cognitive and functional aspects. The languages of uncertainty and quality in science must be sufficiently clear and robust that they can be easily learned and shared among different sorts of experts, and also among the different sorts of inexpert publics, for dialogue and debate. Finally, now that uncertainty is politicized, this new languages of science must incorporate the complementarity of objective knowledge and value-driven commitments. This last requirement is to be fulfilled by the language being designed to reflect the dialectical interaction of knowledge and ignorance.

8.1. NUSAP: DESIGN CRITERIA

Notations, like other graphic representations, are the product of a design process. We do not know the sorts of considerations which went into the shaping of notations in the past; only the completed symbols themselves survive in the historical record. For the NUSAP notational scheme, we can describe explicitly the design choices that were made, in order to best fit the notation to its functions. We may now define a set of criteria in whose terms NUSAP is a good design solution, in the context of its intended functions. These are as follows:

– *Simplicity*. This criterion relates to the signs themselves, to be written and read by users with all kinds and degrees of scientific expertise. All the categories of NUSAP are represented by familiar symbols in a simple string. This feature is of the greatest importance for communication.
– *Naturalness*. As an extension of existing usages, this criterion enables the assimilation of NUSAP to familiar notations. Thus the first three places represent completely standard ideas; the fourth occurs frequently in statistical practice; and only the last of the five is radically new. A NUSAP expression may appear unfamiliar at first; but since it is such a natural extension of existing usages, it should present few problems, typographical or conceptual. This aspect is very important for the diffusion of a new notational system.
– *Consistency*. The scheme must be consistent with the whole range of existing operations using arithmetical language. This includes counting, measuring, indexing, naming and estimating; indeed, the scheme lends greater transparency to all such operations, by making explicit their underlying assumptions and their relations to each other. Through NUSAP, the interactions between counting, estimating and measuring, which we described in an earlier chapter become clear and easily understood. The scheme can also be used to elucidate the intended meaning of quantitative statements expressed in ordinary language, as one-in-a-million.
– *Flexibility*. Since NUSAP must be widely diffused in order to perform its functions, it must possess flexibility for communication in a wide variety of contexts. Policy-related research spans the whole range of activities from the most technical of investigations to the most political of debates. A notational scheme that can carry information along such a range (with appropriate changes *en route*), will be of great advantage in the policy process. A subtler form of flexibility involves the ability to express closely related forms of a single quantitative statement. In this way, different nuances may be conveyed, appropriate to the messages intended for particular functions.

- *Clarity*. This criterion ensures that NUSAP expresses what is best in the traditional craft practices of quantification, so that these are easily understood and can then be used to guide developments in new applications. By fostering awareness, it should guide and enhance craft skills, and not attempt to replace them by abstract and opague formalisms. The achievement of a harmony among the different sorts of uncertainty in a single arithmetical expression can now be accomplished explicitly, using the NUSAP categories as guidelines.

For the special functions of NUSAP, four supplementary criteria are important. Indeed, we may asses any notational system for the expression of uncertainty and quality in terms of how well it fulfils these latter special criteria.

- *Complementarity*. This criterion is a reminder of the interaction between the quantitative and qualitative aspects of any numerical expression. It is essential for a good notation to discourage the separation of the quantitative from the qualitative aspects of information. It should foster the habit of conceiving them as an inseparable complementary pair.
- *Demarcation*. This criterion represents a new version of a classic problem in the philosophy of science: which scientific statements (or fields) are genuine, and which are empty, spurious or even pseudo-scientific. The problem in its present form derives from the seventeenth century philosophers and can be traced through Hume, the Vienna Circle and Popper. Nowhere in this tradition is there a recognition that statements may be of an empirical form, and cast in mathematical language, yet pseudo-scientific. Our recent experience of some mathematical social and decision sciences and policy-related research shows that such a possibility is all too real; we have mentioned this aspect in connection with our definition of GIGO-Science. NUSAP can be used in the identification of hyper- and pseudo-quantification, through its integrated treatment of the quantitative and qualitative parts of an expression.
- *Intersubjectivity*. A notational system which depends on skilled practitioners must be carefully designed so that their shared experience can be reflected in its results. It should prevent arbitrary or subjective opinions from being presented as a communal, intersubjective consensus. Thus protected from misinterpretation, evaluations of uncertainty can become part of a constructively critical dialogue, like science itself.
- *Quality assurance*. This will be a deeper enquiry, and a more comprehensive judgement, than is implied by the simple quality-control operation. For the latter refers to the product as tested on completion; the full assurance of quality also involves an analysis of the whole process of production and control, so that future products will be of guaranteed excellence. NUSAP enables such a deeper analysis of information, through its elicitation mechanisms, which are guided mainly by the pedigree category.

Thus NUSAP can contribute to the identification and elimination of low quality information, a task requiring the participation of producers and users alike. Like any instrument of reform that can injure vested interests, NUSAP is vulnerable to abuse. We would be naive to attempt to design it so as to prevent any possibility of corruption. As all designers of systems know, such an over-design not only leads to an infinite regress, but actually opens up new possibilities of error and abuse. Our design of the scheme will not prevent abuse, but will at least require a great effort in the production of a plausible counterfeit NUSAP expression; and given an aware audience, this would still be at a high risk of exposure. Thus, while not preventing the corruption of information, NUSAP can at least inhibit its practice and better expose it to public view.

8.2. NUSAP: PLACE-VALUE

In our earlier discussion of design, we considered the place-value-with-zero scheme for ordinary numbers as a leading example of conceptual design. Such a scheme is of great generality and flexibility. We recall that it comprehends all number bases from two upwards; the places refer to powers of the base; independently of what the base may be. Once a base is chosen, particular sets of digits may be entered, giving a representation for particular numbers. Any arithmetical operations, either calculation or comparison, must be done within the same base, or as we call it, notation. Otherwise nonsense would result, on the mixing of bases.

The NUSAP scheme has strong affinities with the place-value representation of numbers. In its form, it is a string of places (or boxes) called "categories". These are related to NUSAP acronym, being *Numeral, Unit, Spread, Assessment* and *Pedigree* respectively. It can be seen also as a string of schematic letters

Categories	Numeral	Unit	Spread	Assessment	Pedigree
Schematic Letter	N	U	S	A	P

The categories operate implicitly to provide meaning to the symbols of the string; just as the places in the digital place-value system define that power of the base by which the digit is to be multiplied. Thus an entry in the third place of NUSAP will represent a spread, as one in the third place (from the right) in a numeral will be multiplied by the square of the base. Of course, the places in NUSAP have qualitative differences in meaning, not merely quantitative differences in size; and the entries can be very varied in character, unlike the digits in numerals. This is one reason why we put an explicit sign of separation between the places in the string; for convenience we choose the colon(:) and we read it as "on". The schematic representation of NUSAP will be N :U :S :A :P.

Just as we can usefully speak of the place-value-with-zero scheme, and then of particular choices of base, and finally of particular instances, we can make a similar description of NUSAP. At the most general level we have the *scheme*, consisting of the five schematic letters and their implicit meanings. Next we assign the five sets of symbols, standing for concepts and operations, which may be entered to replace the schematic letters in the string. Each such fixed collection of the five sets, we call a *notation*. Finally, elements of each of these sets may be entered in the places of the string. When this is done, the scheme yields a fully meaningful quantitative statement called an *instance*. We can illustrate this logical structure of NUSAP, and compare it to the place-value scheme for numerals, using only the first three places for the moment, by the following example:

	NUSAP			Numbers		
Scheme (powers)	N	U	S	2	1	0
Notation (base)	**R**	MKS	$\pm n\%$	10	10	10
Instance (digits)	0.2	m $\cdot$ sec^{-1}	$\pm 2\%$	4	5	0

The meaning of each entry of an instance is given by its place in the string, and by the chosen notation for that place as defined by the scheme. Thus the 5 in the second place of 450 (instance) is evaluated by multiplication with 10 (notation) to the first power (scheme). Similarly in the NUSAP example, the symbols m.sec^{-1} in the second place of 0.2: m.sec^{-1}: $\pm 2\%$ (instance), are within the MKS system of measurement (notation), filling the unit place in the NUSAP scheme. In the other two places, **R** refers to the ordinary set of numbers, and $\pm n\%$ an integer percentage spread.

Another qualitative difference between NUSAP and place-value is the nature of the progression through the places. In the case of numerals, only the order of magnitude of the multiplier changes. In NUSAP, we notice how the front end of the scheme is more quantitative and familiar. Indeed, an instance for the three first categories can quite well *appear* identical to one in the ordinary use, except for the colon signs (if they are used). In fact, with some notations for the fourth NUSAP category, assessment, the similarity persists, leaving only the fifth category, pedigree, as anomalous or novel. However, as the complete NUSAP representation of five places fills up, the initial similarity is almost always dispelled.

We may well ask, why bother all that logical apparatus and special extra symbolism, only for the sake of one extra qualifier onto a numerical expression? The first answer is, that an extra bit tacked on to notational systems which are not particularly coherent with themselves or standardized, would only add to the prevailing confusion. But there is a more positive and substantive reason as well. NUSAP is not merely a collection of symbols and notations, it is a logically coherent system. In its form, it shows the *complementarity* mentioned in the design criteria; the categories are closely related one to the next, and also move continuously from the "hard" numeral to the "soft" pedigree.

Also, only a logically coherent system can act effectively for the criteria of *demarcation*. A well formed quantitative expression in NUSAP will be the result of a process of reflection on the production of the result, and of harmonization among the entries in the notation for the different categories. Quantitative statements which, on analysis in NUSAP show themselves incapable of such harmonious representation, are exposed thereby as being weak or negligible in content.

For the *assurance of quality* as well, the use of the full NUSAP system is essential. Those who use it come to see quantitative expressions in a "NUSAP" way. On a NUSAP string, any blank spaces, (which in ordinary notational systems do not occur) call for an entry. We cannot remain ignorant of our ignorance concerning the qualitative aspects of quantitative statements. Equally to the point, NUSAP protects others from remainin such a state of compounded ignorance about our quantitative statements.

With the NUSAP approach, a quantitative expression, even if written informally without colons, will never look the same. For, to fix the degree of precision of the numeral entry (which may range from several significant digits to an order-to-magnitude), we automatically consider not merely the given spread, but also the more qualifying parts as well. An expression with excess precision in the numeral place will instantly strike the NUSAP-trained eye as inharmonious and the product of unskilled work. Thus the formation of a NUSAP expression may require a sort of "tuning-up" exercise, where the different categories of the scheme are scrutinized and brought into harmony. Once the cognitive uncertainties are well expressed, there may be a further "tuning", so that the expression may best perform its functions as an input to a particular policy process.

Thus NUSAP becomes a total way of perceiving, conceiving and using quantitative information. Its categories provide stimulus and guidance for reflection on numerical expressions and for their analysis. We will show how NUSAP's procedures of elicitation bring out their structural features. Once these procedures are learned, they provide a framework for a skilled and sensitive appreciation of quantities as products of human design and art. The NUSAP techniques also enable the practical realizations of a philosophical point of some importance: that there is no unique representation of a quantitative statement. Rather, there is a family, with meanings overlapping more or less strongly, with different representations appropriate for different circumstances. An easy analogy is provided by place-value, where the same number may be represented by many notations, each on a different base. A more subtle example is when a number, in place-value, is given in aggregated form, as when some zeroes are fillers rather than counters; here the meanings may be quite different. A more significant example of the same point within mathematics, is the use of various special functions to represent a single known number. For example, a particular number can be expressed either in decimal or as the product of its prime factors, or (if it is of that form) as a factorial. Thus 120 is $5 \times 3 \times 2^3$ or $5!$; the choice of representation depends on the context and function. To write 120 when $5!$ is called for would most likely destroy the flow of the argument, and would be taken as a sign of mathematical illiteracy.

8.3. TRADING–OFF UNCERTAINTIES

We have already mentioned the need to "tune" the different categories in a NUSAP expression. How this is to be done, will depend strongly on the intended use, including the nature of the message to be conveyed. An example of this process which is familiar in many practical contexts is the operation of widening the interval of the range of a parameter beyond its reported value, hoping that its unknown true value is more likely to be captured thereby. Thus if we have been given a report of some quantity with a best estimate of 3 to within a factor of 2, but we have reasons to distrust the source, we may simply make the range a factor of 4, for "safety". In NUSAP terms, this involves a trade-off between spread and assessment. That is, letting the initial expression be 3: U: f2: low we modify it to 3: U: f4: medium. This may be seen as the traditional "Roman wall" principle for safety: if you are afraid that it might fall down, make it thicker.

The pair of NUSAP expressions shows clearly what is going on; they represent not quite "the same" number, but slightly different quantitative statements based on the same input data. What is being traded-off here is the claimed degree of exactness of the numerical expression. An increase is spread represents an increase in inexactness; one in assessment represents an increase in confidence, or assessed reliability. The two categories express opposed evaluations of their respective aspects of the quantitative information, and so the trade-off is represented by an increase in both.

With NUSAP we see clearly how the skill of the researcher is used in setting the terms of the trade-off. Thus if it is deemed supremely important to prevent an erroneous number being carried through a calculation or reported, the expression might be modified to 3: U: f100: high. But this statement approaches triviality: why not write 3: U: f10^9: total? In such an expression, we are in effect saying that an estimate that includes all conceivable numbers is perfectly reliable: it must include the true value. But this certainty is bought at a very high price; in a sense the statement nearly becomes true by definition. It cannot be refuted or (more generally) tested; and it has no empirical content. In this way the expression conveys a pseudo-empirical statement, empty of practical meaning. Its perfect assessment is analogous to the certainty of a formula of pure mathematics, as $1 \leq 3 \leq 9$. By enabling such distinctions as these to be expressed simply and clearly, NUSAP performs its functions as a means of demarcation.

Another variety of the practice of trade-off is the use of an "adjustment" factor on statistical information, to compensate for known under-estimation of some relevant phenomenon. This happens most notoriously in criminal statistics, but it also occurs in health statistics and elsewhere. In Britain, the issue of food poisoning become politically sensitive in early 1989. The public learned that the total annual number of reported cases of salmonella infection was estimated from medical data as 20,000; everyone seemed to agree that this number was an under-estimation. Some experts accepted an American figure of 1% for the "reporting rate"; others claimed that 10% was a more accurate proportion. Unfortunately, instead of these alternatives being presented as a trade-off inexactness and confidence (or spread and assessment in NUSAP terms), they were offered to the public as factual assertions of the true number: either a massive 2,000,000 cases or "only" 200,000. The argument about this unknowable adjustment factor diverted attention from the real issue, that there seemed to be a serious problem of health and of regulation in the country, regardless of which estimate is chosen. (We recall our earlier discussion of deforestation in the Himalayas, agreed to be serious in spite of enormous differences among estimations).

Here we have an interesting contrast to the previous case of trade-off. If we interpret the adjustment factor as defining the upper limit of a range of inexactness (as expressed in spread), we find that the quality of the information as an input for policy purposes does *not* decrease as the spread goes up. In the previous case, the concern was to avoid a false statement; that is one where the "true" value lies outside the range of the estimate. The solution there was to increase the stated range, and thereby also the confidence, but at the sacrifice of empirical content. In this example, the true value is not the issue; it is enough that it lies within a range, defined by a *lower* bound that is large enough to indicate a serious policy problem. Hence an increase in the upper bound of the spread does not adversely affect the functional quality of the information. Of course, when the necessarily speculative adjustment factors are presented as facts, the supposed size of the epidemic does become politically significant, but not in a fruitful way.

By means of these examples, necessarily argued in a schematic way, we see how the NUSAP approach fulfils our design criteria. Only within such a coherent framework could we analyze these issues in a clear and disciplined way.

CHAPTER 9

THE NUSAP CATEGORIES: NUMERAL, UNIT AND SPREAD

9.1. NUMERAL

We now consider the NUSAP categories in the order of their appearance. The most quantitative, and also the one which most basically defines the expression as a whole, is numeral. A notation for numeral will express an "arithmetic": a set of elements and the relations among them, which correspond to analogous operations in the empirical system under consideration. The simplest example, of course, is the set of integers and its various extensions (such as fractions and negative numbers, and quantities like $\sqrt{2}$ and π). We have seen that even in these ordinary number systems, the appropriate set of arithmetical relations may be artefactual or non-standard.

The elements of a notation for numeral may well be less precise than numbers. If a quantity is known only as an interval which lacks any preferred point of likelihood or of symmetry, then this could be the entry in the numeral place. Thus we could have (a,b): for an ordinary interval; $(\geq a)$: for an open-ended one. We may also use expressions of yet more general mathematical structure, as a finite set of integers representing an ordinal scale (say -3 to $+3$) as in much of social research; or numbers representing indices or scores with a purely artefactual arithmetic.

As these examples show, the numeral place can be used to express a strong uncertainty. Indeed, the balance of uncertainties displayed in the numeral and spread places is an important element in the harmonious design of a completed NUSAP expression. In this regard, the notations for numeral convey a topology, analogous to that which we have discussed in connection with maps.

9.2. UNIT

Unit expresses the relation between the arithmetic of the numeral place, and its application to practice. It is the link between a number system which would otherwise be uninterpreted, and the empirical world which is relevant in the quantitative expression.

It is common in scientific practice for units to be expressed with a quantitative prefix, as *milli*meter or *kilo*gram. The use of such prefixes is a consequence of the various constraints of systematic measurement and of its description. It is frequently, or indeed usually, necessary to do measuring or counting on a scale which is very different from the "fundamental" one. Thus ordinary rulers are divided in the sub-unit millimeters, and that is the practical unit for measuring at that scale; while long distances are measured in terms of the aggregated unit kilometer. There is a similar effect in counting; thus the aggregated unit £K appears in financial reporting as the effective unit, even if the sum was derived from traditional accounting procedures with a precision given in pence.

These two aspects of unit involve different sorts of uncertainty, and therefore within this NUSAP category we distinguish between the *standard* and the *multiplier*. By their means we can distinguish among the meanings of terms which may be equivalent in an abstract arithmetical sense, but whose particular meaning, and associated uncertainties, are very different. Thus in the examples we used above, "milli" is a multiplier on the standard "meter". But "kilo" is not a multiplier on the standard "gram" in the Systeme International (S.I.); for "kilogram" is the fundamental unit there. At a more practical level, we observe that the various units of energy as Watts, Kilowatts, Megawatts and Gigawatts are not simply the same standard unit with a set of different multipliers increasing by a factor of a thousand, corresponding to the prefixes. Rather they refer to physical and accounting operations at very different levels, and relate to very different policy uses. Thus, writing in the NUSAP notation we do *not* have $1 : Kw = 10^{-6} : Gw$, except in a purely formal arithmetical sense.

The simple design feature of distinguishing between standard and multiplier serves mainly for clarifying questions concerning the scale in which operations are performed. By itself it is not adequate for the representation of the complex and paradoxical properties of measurement which we described in connection with the work of N. R. Campbell. However, the issues raised by an elicitation of standard and multiplier within the unit category, can at least make users aware of aspects of measurement which are not entirely straightforward. Thus, even in connection with unit, the NUSAP system fosters an awareness of deeper sorts of uncertainty, as they relate to reflective and self-critical practice.

9.3. NUMERAL AND UNIT: NUANCES OF EXPRESSION

A few examples will show how the flexibility of NUSAP in forming and combining numeral and unit notations enables the expression of quantitative representations which are both clear and nuanced.

We begin by observing that the number "five million" has a variety of representations. With the awareness provided by NUSAP it should not take us long to query the meaning of the common form 5,000,000; we should immediately ask, how many of those zero digits are fillers; and, correspondingly, whether the aggregated unit is million or some other base. Moving on the other representations, we have the more scientific notations 5×10^6 or 5E6. These seem unambiguous, so that the only appropriate NUSAP representation would be $5 : 10^6$ or $5 : E6$. However, consideration of an apparently similar case shows unexpected possibilities of ambiguity and nuance. We recall "the billion story"; now we consider "the million". Among the alternative meanings, we may start with the integer that lies between 999,999 and 1,000,001. Moving on, we have 1 : E6 or perhaps 1,000 : E3 (or more clearly 1K : E3). But the term "a million" may not be intended to denote a determinate numeral; frequently such a term appears in estimates, functioning as an expression of an order-of-magnitude. For instance, in the theory of acceptability of risks, the baseline one-in-a-million is frequently invoked as a magic number. Thus "a million" may express a much coarser topology than, say 1 : E6. To escape from the pseudo-precision conveyed by the 1 in the numeral place, we may put a filler symbol there, writing, for example, – : E6 to express the order of a million.

All these notations presuppose that the name million conveys no special meaning of its own; that a million means precisely the same as 10^6 or E6. But these expressions tend to be used in different contexts; while scientists or computer experts will frequently name the quantity in speech as ten-to-the-six or E-six, still the locution million survives in common speech. Such variants all belong to the same family of meanings; but it is clear that the circumstances under which each of them is appropriate are not identical. Thus the mathematical notations explicitly provide information on the relation of million to the unity and (by implication) to the other aggregated units: the 6 is explicit. In the prose form, it is not there and "million" can function as the unit of generalized large quantity, like "myriad" in Greek. This is most clear in connection with its "magic number" function as in risks. Indeed, there the mathematical forms as E6 are decidedly inappropriate, in view of the severe uncertainties in all such estimates. Hence for a full range of expressions for this aggregated unit of counting, there should be a notation for "million" itself, perhaps M standing alone. This symbol is already in use as the "Mega" prefix; and as we remarked, this is not simply a mere thousandfold multiple of K for kilo, but relating to a different set of meanings and uses. When we pass to billion and trillion, and beyond, it becomes clearer that their rhetorical "magic number" functions are at least as important as the more strictly quantitative representation.

These insights can help us in the naming of quantities which have less ambiguity than million alone. For example, half-a-million is not the same as five hundred thousand or even 0.5 million with its implied high precision. In NUSAP, we could write $^1/_2$: M, 500 : K and 0.5 : M respectively. The first form reflects the sorts of estimation that occur in practice. The second can be a real aggregated count; this conveys greater precision than the estimated $^1/_2$: M. Finally, 0.5 : M seems unlikely to reflect practice, although 0.5 : E6 can be the result of a calculation. The use of fractions in such contexts is an innovation. In ordinary place-value notation it is impossible; but the separation of places for numeral and unit in NUSAP enables it to occur naturally. Thus we can write $^1/_3$: M or $^1/_3$: E6 as the rough cutting of the aggregated unit; this is much more natural and correct than, say 0.33 : E6.

Another useful example from everyday life is counting in dozens; this shows more clearly the influence of the process of production of the quantity, since in this mixed numerical base there is no ambiguity between counter and filler zero digits. Thus for a quantitative statement about eggs, we may have as alternative representations $4^1/_2$: doz-eggs, 9 : half-doz-eggs and 54 : eggs. The first corresponds to the price, the second to the packaging, and the third to the use (in small quantities) in cooking. This example illustrates how an expression that may be appropriate in one context (integer counts for use) may become pseudo-precise in another (prices or packaging). Using large units of aggregation which have been traditional in the retail trade, we can again make the distinction between 72 eggs, 6 dozen and $^1/_2$ gross.

In addition, the ambiguities of ordinary digital notations, as exemplified in the "fossils joke" do not long survive critical scrutiny in NUSAP terms. Thus, if we are to make a sum of, say, a million and eighteen thousand, we would have

$$
\begin{array}{ll}
1,000,000 \quad \text{or in NUSAP} & 1 : E6 \\
+ \quad 18,000 & + 18 : E3 \\
\hline
\end{array}
$$

Formally, the arithmetical example sums to 1,018,000; and it is left to the personal interpretation of the user, what meaning this total has, if any. By contrast, in the NUSAP form, the sum is not possible, since the units are different, even though they are instances of the same notation En. We must then decide, whether to write 1 : E6 as, say, 1,000 : E3; or alternatively, 18 : E3 as 0.018 : E6. Either way, the choice depends on an interpretation, for which there should be some grounds. These will reflect the operations whereby the two terms were obtained. Depending on the outcome of this analysis, we may write the sum as 1.018 : E6 or 1,018 : E3, or alternatively, simply 1 : E6. For this case reflects a situation where the larger term of the sum swamps the smaller. Thus NUSAP enables us to move beyond the simple distinction between counter and filler digits.

The NUSAP approach provides a means for guiding the perception of quantitative expressions. The availability of so many alternative NUSAP notations enlarges the imagination; new possibilites are suggested, and new questions about meaning, balance and appropriateness can be raised. In this way, the information content of quantitative expressions (even with only numeral and unit) can be significantly enhanced. This improvement does not require special mathematical symbolisms that impose barriers to inexpert readers.

9.4. SPREAD

The category spread lies in the middle of the NUSAP scheme in several senses. It is the first of the categories to qualify others in relation to their uncertainties; with it, the qualitative or "soft" part of the five-place scheme begins. Spread expresses the simplest form of uncertainty, inexactness, as in the traditional statistical variance or experimental random error. This relates to the product of the operation (be it of measurement or counting), rather than to deeper uncertainties in the process. In this way it reflects uncertainty at a technical level; whereas the later categories in the NUSAP scheme will relate to methodological and epistemological levels of uncertainty.

By including the spread category in a NUSAP instance we can express even more clearly the points already made about artefactual arithmetics. Thus in the previous example, when we write an expression for "a million", we need to consider how the spread place is to be filled. Ordinarily, 1 : E6 would have a spread of E5, the next significant digit, unless we mean − : E6, an order of a million, or something yet more vague as − : M. Then if we write the sum (1 : E6 : E5) + (18 : E3) we see that the lesser term is not merely very small in comparison to the greater; it is actually swamped by its inexactness. By any rule of artefactual arithmetic involving inexactness, we then have, rigorously

$$
\begin{array}{r}
1 : E6 : E5 \\
+\ 18 : E3 \\
\hline
1 : E6 : E5
\end{array}
$$

With the help of the spread category we can handle cases which would otherwise be ambiguous. Suppose that the smaller term is 180,000. Then if we add (1 : E6 : E5) + (1.8 : E5) we can retain the lesser term and write 1.2 : E6 : E5, where the last digit is uncertain but not meaningless. However, if the sum involves terms which are very large compared to E6, then the spread can be taken as E6 itself, and the 1.8 : E5 may again be swamped. Thus if we have

$$
\begin{array}{r}
156 : E6 \\
+\ 72 : E6 \\
18 : E5
\end{array}
\quad = \quad
\begin{array}{r}
156\ \ : E6 : E6 \\
+\ 72\ \ : E6 : E6 \\
0.18 : E6 : E4 \\
\hline
228\ \ : E6 : E6
\end{array}
$$

In this latter case, the large entries in numeral indicate that these are some sort of counts, and that there is no interpolation within the scaling; hence the spread is E6 itself.

The flexibility of NUSAP enables the development of appropriate new notations as need arises. As we have seen, the quantities involved in policy-related research are frequently so inexact that the traditional " ± " notation is inadequate. For such cases, we offer the term "within a logarithmic interval of n", for example, "within a logarithmic interval of 10". This expression is better than the one in common use, "to within a factor of 10", for this is ambiguous, possibly referring to factor 10 on both sides of the number, leading to a logarithmic interval of 100, or perhaps referring to the logarithmic interval of 10 itself. It would be pedantic and pseudo-precise to express the end-points of the logarithmic interval as $(10)^{1/2}$ more or less (multiplicatively) than the given quantity: a rough approximation, giving a convenient digit, is appropriate. For $\sqrt{10}$ we can use 3, always rounding to the nearest illustrative number. Thus 3 : U : Λ10 could be represented by the inequality $1 < x < 10$, while 5 : U : Λ10 could be $1\frac{1}{2} < x < 15$. (Here U is the schematic letter corresponding to the Unit category).

In the policy context, fractions less than the unity, expressed as percentages, are frequently used to indicate the division of some aggregate. The inexactness of such estimates is extremely difficult to represent in a compact notation, and a misleading impression of precision is all too often conveyed. Thus 40% may mean "less than half but more than one third", or perhaps "less than half but more than one quarter". These inexact quantities may be represented in NUSAP as $1/3$: 1 : $< 1/2$ and $1/4$: $< 1/2$ respectively. These forms tend to emphasize the lower bounds; if we wish to emphasize the upper bounds, we may write $1/2$: 1 : $> 1/3$ or $1/2$: $> 1/4$. To stress the interval as a whole, we may use the notation mentioned previously, as $(1/3,1/2)$: U : − or $(1/4,1/2)$: U : −. If we wish to remind users that we are concerned with a particular point-value within that interval, we may write x : U : $(1/3,1/2)$ or x : U : $(1/3,1/2)$. By such means, we can express quite fine distinctions among inexact estimates of fractions, avoiding the hyper-precision inavoidably associated with the two-digit percentage. All these forms enable us to express clearly that the means of production of the quantity do not provide us with information for distinguishing among numerical values within the interval. (Of course, we must not forget that the end-points are themselves vague). We can refer to this interval as an "indifference class", in the sense that no one numerical value can legitimately be taken as a representative of the class in preference to any other. In symbols, we write the general case as x : U : S.

9.5. TOPOLOGY: GRID AND RESOLUTION

These two ideas, of using fractions and of the "indifference class", can be combined to enable a clearer understanding of the use of number systems in the context of uncertainty. They shed new light on the meaning and relations of the numeral and spread categories of the NUSAP scheme. We have in several occasions used the word topology, qualified as more coarse or more fine; by this term we mean certain structural properties of the space presupposed in our quantitative expressions.

For our purposes here, the topology is best understood through the example of the map. There we discussed the precision of the grid of coordinates:how many successive subdivisions of the basic scale are employed. We also discussed the "resolution", that is the minimum distance between objects for them to be distinct on the map, or the minimum size of an object for it to be represented. These two aspects of a map, grid and resolution, must be harmonized for a good design. As we mentioned, a motoring map needs to convey information at a glance, and so should not not be clattered with detail, hence its resolution will be less fine than a topographical map of the same scale and grid. For another example, a computer VDU picture done in large pixels, portrays a resolution that is identically equal to grid.

These distinctions are built into the NUSAP scheme, where grid is normally expressed by numeral and resolution by spread. The above example of percentage estimates can be interpreted in terms of the design weaknesses of the decimal place-value system for grid as well as for resolution. The complete precision of a digit (such as 4) as a counter, creates an inevitable hyper-precision when strings of such digits are used for measurement or estimation. (For this we may blame the system of numbers if we are empiricists, or we may blame the imperfect physical reality if we are Platonists).

We have indicated that fractions possess a flexibility of expression, avoiding hyper-precision. Moreover, they enable the user to control the grid to best advantage. Instead of the grid being constrained to cascade down by factors of ten, its last place has a great variety available. It can be in halves, thirds, quarters, perhaps even fifths (or some combination thereof). How far in the partition one chooses to go will depend partly on what the user can grasp (we might use sixths or eights for a pie-chart); and also on the fineness of discrimination which is justified by the production process (including instruments and theories). Thus the consideration that explicitly make up the last category of a NUSAP notation, pedigree, enter into the design of the first, numeral. In some fields of science, like biology, it is permissible to express the "last digit", the uncertain place, by means of a fraction; thus, for example, $3.2^{1/3}$. This extension of mixed fractions from integers to decimals seems eminently sensible. The formally equivalent expression 3.23... conveys a different meaning, patently hyper-precise in this context.

Now we consider resolution, which is normally expressed in spread. This corresponds to the "indifference class" mentioned above: the set in which elements cannot be distinguished. This class may relate to a particular number, usually implied by its form. Thus $1/3$ can mean any quantity which is significantly more than $1/4$ and significantly less than $1/2$. By contrast, 0.3... stands for the infinitely recurring decimal which with mathematically perfect precision defines a real number. This form implies that it is different from a number which is all 3's except for a 2 in the thousandth place. Such precision is totally inappropriate outside pure mathematics, and yet it is forced by a digital notation with its implied fine resolutions.

It is not merely fractions and decimals that show differences in implied resolution. We have already emphasized the ambiguity of the zero; and clearly the aggregation of units of counting is equivalent to changing resolution at this larger scale. Some particular numbers convey a special resolution. Thus 5, lying halfway between 1 and 10, may convey the more broad resolution appropriate to $1/2$, rather than the narrow band of the number lying between 4 and 6. The digits 3 and 6 operate similarly, although to a lesser degree. In large numbers, the initial digit 1 expresses a vagueness reminiscent of the prefix in "a million"; and so 1 : E6 is not identical to $1/2 \times 2$E6. Appreciation of such nuances enables a more clear and powerful expression of quantitative information, without needing to invent or invoke special symbolisms or conventions. In such ways the NUSAP approach can improve the communication of quantitative information even when it is only in the general background understanding.

We can now clearly see the insufficiency of the significant digits conventions. They are an *ad hoc* device for overcoming the inappropriateness of a system for new functions. We recall that decimal fractions were invented by Stevin. They enabled the extension of a system designed for counting discrete quantities, onto the domain of the representation of continuous magnitude. Problems of resolution are minor in ordinary counting, but become important in measuring. In relation to the function of managing uncertainties the rigid set of cascades by ten is a design defect of the system of decimal fractions. This is seen most easily when we want to convert between one set of units and another, say Imperial and S.I. Thus, an inch equals 25.4 mm, and a tolerance of 0.01 inch corresponds to that of 0.25 mm. There is no way that significant digits, cascading by 10, can translate tolerances between different scales; hence there must be special conventions, which may be cumbersome or even misleading, for the translation of tolerances between systems. This point is *not* a criticism of the decimal system, it is merely an observation that its design has defects for certain functions, most notably those associated with un-certainty.

The indifference classes illustrate some very fundamental contradictory features of the application of mathematics to our empirically measured reality. We have discussed Campbell's paradox of overlapping tolerances (or indifference classes). This has the effect of rendering ordinary arithmetic, with the axiom of transitivity for the equality, incongruous with the operations of measurement. For we may have $A = B$ and $B = C$, both relating to over-lapping indifference classes, and yet $A \neq C$. Arguments involving the manipulation of indifference classes go back to the very earliest recorded human history. When Abraham argued with God about the proposed destruction of Sodom and Gomorrah because of their sins, he first obtained a target of 50 righteous men, which would be sufficient for them to be spared. Then he secured an indifference class around that ("Would you destroy the city for a mere 1 or 2 less than 50?), thereby bringing the target down to 45. By iteration, this first Jewish argument quickly reduced the target of the righteous to 10; but in the event, even that modest requirement could not be fulfilled.

Recognition of grid and resolution is provided by a very simple elicitation procedures. This may be used to guide an introductory enquiry into the appropriateness of quantitative expressions, and also to enhance the skill of such enquiry. This elicitation has two parts, one relating to cascades and the other to digits. Suppose we are presented with a number, say 4.32; we first ask, "do we have too many digits here?" This is a fairly straightforward and familiar question. Then we ask, "too few?" This recalls our point about hazardous wastes; since measurements to the nearest tonne were meaningless, why not measure to the nearest kilo? The absurdity of such an increased refinement of grid can emphasize the possible inappropriateness of the grid as it exists.

Complementary to this pair of questions are those relating to the last digit, or resolution. We may ask, given 4.32, can we justify, in terms of their production, the distinction between this and either 4.31 or 4.33? If either answer is negative, we iterate, always relating back to the original number to avoid the drift of Campbell's paradox. In this way we build up an indifference class pinned to 4.32. How to represent this class, given the constraints of the decimal system, is a design task. If possible, a fraction can be used in the last place, or perhaps an expression for spread can be added on. Further complications arise when the number is combined with others, as in the case of the π-dilemma. Thus there remain some challenging design exercises, even in this apparently elementary field.

9.6. SPREAD AS A QUASI-QUANTITY

Spread is not a quantity in the ordinary sense, for except in cases where it is derived in a highly mathematical statistical exercise, it does not itself have a spread. This is not an issue of idle philosophical curiosity, for we have seen that recursion on mathematical measures of uncertainty (as in the case of Bayesian probabilities and fuzzy sets) can lead to an infinite regress and the associated insoluble problems of interpretation and practice. We do best to think of spread as an estimate, whose appropriate arithmetic (if it must be formalized) has a very coarse topology. With spread we encounter the contradiction inherent in all digital representations: even when we wish to express vagueness, our digital symbolism prevents this. In this case there is no question of an infinite regress, for spread is (in conjunction with numeral and unit) adequately qualified in the NUSAP scheme, by assessment and pedigree.

Thus, proceeding through the NUSAP categories, we not only move from the quantitative to the qualitative, with spread on the bridge. We also move through types of reference, where the latter categories (including the last two as a pair) qualify the preceeding ones. Such a complex conceptual structure frees the NUSAP system from the contradictions afflicting the attempts at formalization of uncertainty by means of a single calculus. Such unitary formalisms embody the dream of a single philosophical language in which all problems can be framed and solved, an aspiration found in such diverse figures as Lull, Leibniz, Russell and Carnap. We do not believe in the separation of the formal and the informal, and the exclusion of the latter as representing a lesser form of knowledge, analogous to "knowing-how". The NUSAP approach includes the less formal with the more formal, as well as the more qualitative with the more quantitative, in a continuum. The basic distinctions are not blurred or obliterated; they remain as polar elements in a set of complementary pairs in a dialectical unity. All this analysis is evoked by reflection on the spread category, itself a generalization of the simplest sort of uncertainty in quantitative information.

CHAPTER 10

THE NUSAP CATEGORIES : ASSESSMENT AND PEDIGREE

The final pair of categories, assessment and pedigree, are the most obviously innovative features of the NUSAP scheme. They represent levels of uncertainty that go beyond those which can be managed by purely technical means. As such, they are the most qualitative categories in the scheme, and also the most flexible in their possible interpretations and associated notations.

They express conceptual operations of criticisms and evaluation on the entries in previous categories. They are reflective, and they foster self awareness among users. The operations related to assessment and pedigree cannot be performed by automatic means, neither by a formal calculus nor by a computer package. Nor can they be accomplished in isolation from the whole body of relevant scientific knowledge and craft skills. In this sense, they introduce philosophical considerations into the practice of research; but this is done in a constructive rather than in a sceptical way. The result of these reflective activities can be operationalized in special notations expressing the distinctions relevant to any particular case. These notations enable the development of conventions for a useful "arithmetic of uncertainty", for the combination of assessment ratings, which are gauges (with a very coarse topology) in the sense discussed in Chapter 7.

We can exhibit the structure of the NUSAP approach by a slight modification of the original string of five places. We separate it into three sections, consisting first of numeral, then unit and spread, and finally assessment and pedigree. The first section corresponds to pure mathematics; that is, an uninterpreted calculus, as common arithmetic standing alone, or algebra. The second section describes the application of the mathematics, as in measurement or estimation. The unit entry relates number to operations through topology and scale, and spread conveys the ever-present technical uncertainty of the operation. Thus, this middle section modifies the first, and conceptually it may be said to include it, since it provides the extension of the arithmetic to a larger system relating to empirical operations.

Finally, the third section of the NUSAP categories is the most general and inclusive, introducing the contexts of production and of use. The former includes the state-of-the-art of the relevant field of scientific production, and thereby conveys the border with ignorance, the deepest sort of uncertainty. The latter, the context of use, includes the policy process in which the quantity is used as an input. It extends even to the general culture in which it is received, and which in the last resort shapes our concepts of science and of knowledge.

10.1. ASSESSMENT

Assessment comes after spread in the string of places in NUSAP; and in its simplest interpretations is closely related to it. In statistical practice, confidence limits will be naturally expressed as an assessment entry. Indeed, in many technical contexts, the assessment category may be considered as nearly an alternative form of spread, and hence redundant. Thus if there is a normal distribution involved in a test, then the confidence limits relate in an entirely straightforward way to the variance; and little extra information may be conveyed by having two entries instead of one.

This close association in an extreme case tends to conceal the radical difference between the two categories. By permitting an impression of similarity between them, it inhibits the understanding of either. Those experienced in statistics are the most prone to make this easy and misleading identification. Even within statistical practice, any but the least demanding examples, illustrate the independence of the two categories. For instance, we may consider a statistical process depending on a number of trials, as a coin-toss or a modelling exercise. Suppose we are interested in the value corresponding to the 95th percentile of the resulting distribution; this will appear in numeral, while assessment will contain the entry $\%95$ (to distinguish the entry from the traditional 95% confidence). The spread entry is then a measure of inexactness, a function of the number of trials. In this way, the meaning of the two terms are independent of each other.

The distinction between spread and assessment is more easily appreciated in terms of the different sorts of "error" encountered in experimental research. We have already mentioned random error, the spread of values obtained through measurement. Contrasted to this is the systematic error, which is estimated in terms of the historic experience of that class of experiments. We have discussed this distinction in the case of the fine-structure constant α^1. Judgements of a still more qualitative character may be expressed concisely in assessment. Particularly in the context of policy-related research, the data may be so sparse and inexact that no meaningful assessment entry can be derived mathematically. Then qualitative judgements are employed on a scale, from "strong" to "weak" or "good" to "poor". For convenience, these may have a numerical code associated with them, understood to be a gauge and not a measure or even an estimate.

The judgements about quality, or reliability, may refer to a broader context than the purely cognitive one of production. If a NUSAP expression is being prepared for a particular use in the policy context, then a pragmatically-oriented interpretation of assessment may be appropriate. The "same" quantity may then be given with two (or more) alternative forms, thus displaying the different meanings which can be conveyed for different functions, by the NUSAP scheme.

10.2. EXAMPLES OF ASSESSMENT

Some simple examples will show how assessment, alone or in combination with the previous categories, can operate flexibly to provide nuanced expressions for quantitative information. For the present we will omit the pedigree category in our discussion of these cases. We may start with the fine-structure constant α^{-1}, which we analyzed in Chapter 5.

In this case a NUSAP notation is

$$x : \text{E--4} : \pm \sigma_{\text{ppm}} : \sigma_{\text{previous}}$$

Here we are using assessment for a concise expression of the historical lesson that the systematic error, as discovered on repeated periodic reviews, tended to be about twice the reported random error of each recommended value. For an instance, we may write

$$\alpha^{-1}(1968) = 137.0 + 360 : \text{E--4} : \pm 1 : \pm 2\tfrac{1}{2} : \text{P}$$

We notice how we can separate off the uncertain part of the quantity, letting the NUSAP expression be an element in a sum whose other term is a fixed quantity, here 137.0.

For an example of a NUSAP representation in policy-related research, we discuss a measure that is of considerable importance of the assessment of the health effects of radioactive fallout, as from the Chernobyl accident. This is, the predicted radiation dose, in Sieverts (Sv), to an individual via the "pasture-cow-milk" pathway following a single uniform unit deposit. This is derived from a simple linear sequential model, involving relatively few parameters (ten). For each of these a probability distribution was constructed based on expert judgment. An uncertainty analysis, with Latin Hypercube Sampling, was used to calculate a probability distribution for the result. Values obtained are 2×10^{-8} Sv at the 50th percentile, with 3×10^{-9} and 8×10^{-8} at 1% and 99% levels respectively.

For a NUSAP representation let us start with unit. This is not simply Sv, for the whole calculation is in 10^{-8} Sv. Hence the unit place is divided into multiplier and standard and reads E--8 Sv. Then the numeral reads 2. For spread we have a factor, from 0.3 to 8, or $\Lambda 5$ (we note that calculated precisely this gives 0.4 to 10; but NUSAP reminds us of the needs of convenience and clarity over hyper-precision). For assessment we qualify the $\Lambda 5$ by quoting the odds 50/1 for the portion of the total distribution contained between the 1% and 99% limits; this fits with the subjective probability distributions used as basis of the Monte Carlo simulations. The NUSAP expression is then

$$2 : \text{E--8 Sv} : \Lambda 5 : 50/1$$

The flexibility of the system enables us to choose the best representation for any quantity, depending on the use to which it will be put. Thus for the quantity 2×10^{-8} Sv, we may be more interested in the process by which it was calculated. Then in assessment we could show that the number is the median, that is at the 50th percentile of the distribution, displayed as $\%50$. In this instance, the spread is the sampling error, depending on the number of runs of the simulation (Funtowicz and Ravetz, 1987b). The relevant NUSAP expression would then be

$$2 : E{-}8 \text{ Sv} : e(N) : \%50$$

We have previously mentioned the prediction of the rise in mean temperature consequent on the greenhouse effect, in connection with the dilemmas of science in the policy process (Chapter 1). The original statement was of a rise between $1.6°C$ and $4.5°C$ over the next forty years. To put this assertion into NUSAP form, our first task is to fix the entries for numeral and spread. The interval $(1.6, 4.5)$ is nicely analyzed into $3 \pm 50\%$. We notice that $\pm 50\%$ yields a logarithmic interval of 3. The end-points for the spread are then $(1^1/_2, 4^1/_2)$, which appropriately reduces the implied precision. For the unit we can write $°C_{2030}$, indicating the physical unit and time together. There is no need for a multiplier, since the issue here is the increase in the standard measure of temperature. For $N : U : S$ we have the forecast increase expressed as $3 : °C_{2030} : \pm 50\%$. Since the original statement did not specify a best-estimate value, it would be better to write the NUSAP expression as $(1^1/_2, 4^1/_2) : °C_{2030} : -$.

Now we come to assessment. Here we can distinguish between two approaches to the question of reliability, and the associated issues of quality. From the cognitive point of view, it is possible to obtain a quantitative assessment rating very simply. The interval $\Lambda 3$ is really very wide indeed; the climatic consequences in this range vary from the bearable to the nearly catastrophic. On a common-sense basis, we may say that the true value is highly likely indeed to lie within that range. However, as we recall from our earlier examples of trade-off, the reliability of the estimate would be even further improved by increasing the spread, say to $\Lambda 10$. In NUSAP form, the forecasts of increase of mean temperature would be

$$3 : °C_{2030} : \Lambda 3 \ : \text{high}$$
$$\text{and} \quad 3 : °C_{2030} : \Lambda 10 : \text{total}$$

Or in interval form

$$(1\tfrac{1}{2}, 4\tfrac{1}{2}) \ : °C_{2030} : - : \text{high}$$
$$\text{and} \quad (1, 10) \ : °C_{2030} : - : \text{total}$$

Here we are reproducing the case mentioned before, where the statement is nearly true by definition; but its scientific quality decreases accordingly. Here the instances of assessment belong to the NUSAP notation which has the qualitative scale: (nil, low, medium, high, total). We shall use the same notation for pragmatic usefulness, below. A scale of such coarseness is appropriate for this sample exercise, with subscripts c and p for "cognitive" and "pragmatic" assessments respectively. We will later exhibit a pedigree matrix for functional quality, from which the "low$_p$" assessment can be derived.

Considering the other approach to reliability or quality, we question the pragmatic usefulness of the estimates as an input to policy. In these examples, usefulness decreases even more sharply than scientific quality. In these terms, appropriate assessment ratings, in a NUSAP expression, would be

$$(1\tfrac{1}{2},4\tfrac{1}{2}) \qquad : {}^{\circ}C_{2030} : - : low_p$$
$$\text{and} \quad (1, 10) \qquad : {}^{\circ}C_{2030} : - : nil_p$$

How could the usefulness of this estimate can be enhanced? For this, we would want to narrow the range of values expressed in spread, so that policy decisions would be based a less indeterminate set of options. In this way, the urgency of the issues could also be better understood. For a narrower range of values to have cognitive quality, the production process for the estimate would need to be reliable in itself. But this is not the case; the whole derivation depends strongly on simulations and guesses. Suppose we considered narrowing the range in spread to $\pm 20\%$; the assessment rating would then drop, say, to low. In NUSAP, we would have

$$(2\tfrac{1}{2},3\tfrac{1}{2}) : {}^{\circ}C_{2030} : - : low_c$$

Can information of low cognitive quality be very useful in such a policy context as this? Hardly likely. Hence the NUSAP expression for this hypothetical forecast, with a pragmatic quality in assessment, would be the same. This analysis shows how in this case there is no easy escape from the dilemma described by *Nature*. With NUSAP, we can at least see clearly what is the shape of the problem. Users are at least made aware of their ignorance, and are protected from what we have called "ignorance[2]". In is in these terms that *Nature* was justified in criticizing the computer simulation as a "cop-out", for that would only transfer the ignorance to a machine. With a clear, unambiguous, and undeniable expression of the low quality of inputs, policy discussion could focus on the consequences of ignorance and the best ways of coping with it, ameliorating it and mitigating its effects. In this way, NUSAP can function to sharpen perceptions of where the problems lie.

For a case where NUSAP can offer significant clarification in a numerical statement, let us return to the report on hazardous wastes. There the raw number provided by the experts was 1,590,014 tonnes (Chapter 4). A first NUSAP expression, evaluating this as it stands, could be

$$1,590,014 : \text{tonnes} : [6 \text{ in } E7] : \text{nil}$$

The spread category expresses the implied precision of the last significant digit; and the assessment conveys the reliability of the statement in the entries for numeral, unit and spread. To improve the quality of this expression, we first split it into its more or less uncertain parts. We recall that we suggested

$$\text{Total Estimated} = 10^6$$
$$\text{Total Reported} = 6 \times 10^5$$

Since any spread in the estimated term will be of the same order of magnitude as the numeral in the "reported" term, there is no point in combining them into a sum. For the first term,, it seems reasonable to suppose that the logarithmic range is 2 or less; this is then a percentage spread of $\pm 33\%$, or $\pm 1/3$. This could have a medium reliability. For the reported entries, the spread could be $\pm 20\%$, a logarithmic range of $3/2$; this we might consider high. Our two expressions would be

Total Estimated 1 : E6 : $\pm 1/3$: medium
Total Reported = $1/2$: E6 : $\pm 20\%$: high

We notice how the instances of unit help to convey the inexactness. The digit 1, even more strongly than 5, avoids the connotations of precision that would be conveyed by, say, 2. This nuance is strengthened by the use of $1/2$ in the second expression; this is of just the right coarseness, compared to 6.

We shall return to some of these examples in connection with pedigree, showing how the full NUSAP representation provides even greater richness and flexibility. How much of the NUSAP formulae should be displayed explicitly, depends on the message to be conveyed. There are some cases when all that is required are numeral and assessment, the latter functioning as a "grade" for the numeral entry. We shall see how his abbreviated form of NUSAP can be useful in some contexts of calculations.

10.3. PEDIGREE

Of the three sorts of uncertainty expressed in NUSAP, ignorance is the most complex, and also the most difficult to convey explicitly. In ordinary scientific practice, ignorance of a special sort is vital to the enterprise: those interesting problems which can be stated, but whose solubility is not assured. These provide the challenge that distinguishes worthwhile research from the pedestrian. In this sense, science deals with controllable ignorance; successful science involves, in the classic formula, "the art of the soluble". Not all ignorance comes in such convenient packages; in contemporary science/technology policy, the most important problems are frequently those of "trans-science" (Weinberg, 1972): problems which can be stated, whose solution can be conceived, but which are unfeasible in practice because of scale of costs. Such trans-science problems may involve ignorance that is quite important in the policy realm, as when decisions must be taken before there is any prospect of the relevant information being produced. In the pedigree category, we do not characterize information (or ignorance) in technical detail. Rather we exhibit the mode of production of the quantitative information being represented, through an evaluative account. This defines the border with ignorance, through a display of what more powerful means were *not* deployed in the production of the information.

Pedigree is the most novel of the elements of the NUSAP scheme; there are no analogies for it in ordinary statistical practice. It provides a guide for the elicitation of the key components of an evaluative account of the information as produced. This is done historically; that is, it is based on experience, rather than on logical distinctions. Thus the concept of pedigree, like assessment, represents a partial formalization analogous to that of chemistry, rather than a complete formalization as in "the calculus". The pedigree expresses the key components by means of a matrix; its columns correspond to the basic aspects (or phases) of the production process; and within each column, the modes express the distinctions which are critical for practice in the relevant field. (By "relevant field" we understand the community of those who are technically competent, as defined in Chapter 4).

The distinctions among the modes of each phase must be ordered in some hierarchy, from the most certain down to the least. When a particular set of modes is chosen to express the pedigree in a NUSAP expression, each of them represents the best possible evaluation in that respect. In this way, the pedigree conveys the state-of-the-art for that item of quantitative information in its field. Thus if we report a "computation model" as the theoretical structure for the information, that implies that there was no "theoretically based model" available, and still less any "established theories", involved in the work. Thus in each phase we are comparing existing results with conceivable alternatives of greater strength. As research fields develop through practice, early pioneering efforts may be superseded by stronger work in such a fashion as this. Hence we may imagine the choice of modes in a pedigree matrix as indicating the border between what is currently feasible and accepted as known, and that which is unfeasible and unknown.

In this respect a pedigree code is analogous to the statement of a proved theorem in mathematics. Such a statement includes more than the result; equally important are the conditions under which it holds. As to other possible conditions, there is ignorance; and the statement of a theorem constitutes an implicit challenge to explore that ignorance. Although quantitative information is not "true" in the same sense as a mathematical result, there is this analogous border between knowledge and ignorance in the specification of its production.

It is very fruitful to look at the modes that are *not* chosen, in particular those lying just above those that are. For these describe a state-of-the-art for that item of information which is *not* achieved; the border between the attainable and the unattainable is defined by the boundaries between the respective modes. These unachieved modes represent degrees and kinds of certainty which are not present in the information as produced. In this sense they correspond to the blank places on the old maps. What sort of information could be achieved, were the process to be improved or transformed, so as to merit higher modes on the elicitation of pedigree, we simply do not know. These might sometimes be conjectures, as the ancient mapmakers depicted their monsters and mermaids; but beyond that border we are ignorant.

This is in its way a negative judgment; the pedigree shows us clearly that of which we are necessarily ignorant. But it has its positive side, in that we see just how far our knowledge extends. Our approach, through pedigree, is complementary to that of Popper. His position was that there can be no perfect certainty in science; and he argued this in general terms, invoking the insoluble "problem of induction". We presuppose the partial and provisional character of all scientific knowledge, and instead of arguing for our ignorance concerning Truth in a general way, we exhibit, in any particular case, the present border between our empirical knowledge and our ignorance.

The existence of such a border depends on there being an agreed and normative hierarchy of modes within each phase of the pedigree. This necessarily derives from an accepted epistemology, in which there is a natural order among all the relevant ways of "knowing". Thus we presuppose that a deductive argument is stronger than an inductive inference; and that this in turn is stronger than an analogical argument; and that all these are stronger than conventional definitions. This last mode permits a wide freedom of choice, and in that sense can be arbitrary and untestable. By contrast, analogical reasoning, while not arbitrary, allows for a multitude of options and interpretations; and its products are still untestable. The product of inductive inferences can be tested and (in principle) falsified; but such inferences do not provide certainty. At the top of the scale is the deductive argument; if it is valid, it carries truth forward, from premises to conclusions, and falsity in the opposite direction. Analogous normative hierarchies hold for the other phases.

10.4. THE PEDIGREE MATRIX FOR RESEARCH

For the evaluative history of the quantity as recorded in the pedigree matrix, we analyse the process into four phases. These indicate, by their various modes, the strength of the different constituents of quantitative information resulting from a research process. We have two cognitive phases, theoretical and empirical; and two social phases encompassing all the sorts of evaluation that we may want to provide.

By including all the phases into the evaluative account of the information, the pedigree achieves a practical resolution of the dichotomy of "knowing-that" and "knowing-how". Our method is analogous to that of a mathematical theorem as stated and proved. Colleagues do not simply accept the statement that the result is true; the study the proof, partly to check the claim and partly to see "how" it is true as the outcome of an argument. The details of the proof also serve to articulate the border with ignorance which is indicated in the conditions of the theorem, and then thereby stimulate and guide further research. The phases in the pedigree matrix reflect the various ways of "knowing-how" the information is produced and accepted; then it is clear in what ways, with what confidence, users will "know-that" it is correct and reliable for them.

The pedigree matrix for research information is displayed in Table IV. In order, the phases are:

TABLE IV
Research-pedigree matrix

Theoretical structures	Data-input	Peer-acceptance	Colleague consensus
Established theory	Experimental data	Total	All but cranks
Theoretically-based model	Historic/ field data	High	All but rebels
Computational model	Calculated data	Medium	Competing schools
Statistical processing	Educated guesses	Low	Embryonic field
Definitions	Uneducated guesses	None	No opinion

- Theoretical Structures
- Data Input
- Peer-Acceptance
- Colleague Consensus

Discussing the separate phases in order, we start with the cognitive phases, of which the first is Theoretical Structures. Following the traditional scientific methodology, we accept that the strongest mode here is *Established Theory*. The general term "established" includes such modalities as: tested and corroborated; or theoretically articulated and coherent with other accepted theories. Thus Einstein's Theory of Relativity was in this sense already "established" when it was tested by the famous astronomical experiment of 1919. When the theoretical component lacks such strength, and is perhaps rudimentary or speculative, then its constructs must be considered as in a "model", but one which is theoretically based; we have then the mode *Theoretically-based Models*. Although still involved in explanation, such a model makes no effective claim to verisimilitude with respect to reality. In this latter respect it is similar to a *Computational Model*, which is some sort of representation of the elements of a mathematical system by which outputs are calculated from inputs. In such a case, there is no serious theoretical articulation of its constructs; the function is purely that of prediction. This mode, Computational Model, characterizes the use of computers for simulations, where real experiments are difficult or expensive; the analysis by Beck (Chapter 2) shows why this mode is not higher in the scale.

Important research can exist where neither articulated constructs nor elaborated calculations is present; this is the case in classic inductive science. Then, with techniques varying from simple comparisons (formalising J. S. Mills' Canons of Induction) through to very sophisticated statistical transformations, we have *Statistical Processing*. Such forms of Theoretical Structure can provide no explanation and only limited prediction; but used in exploratory phases of research, they can yield interesting hypotheses for study. Epidemiological work of all sorts, leading to identification of likely causes of known ill effects, is a good example of this mode. Finally, we have those situations where date which is gathered and analysed is structured only by working *Definitions* that are operationalized through standard routines. This will be the case with field-data, frequently destined for public-use statistics; this is discussed in the next chapter.

The normative ordering among these modes is clear; the higher generally includes the lower as part of their contents. But this does not imply judgements on research craftmanship, pragmatic effectiveness, or on the quality of the investigators or of a field. We do not share in the traditional judgement that all science should be like physics. Therefore, if (in its present state of development) a field can produce only relatively weak results (as gauged by the modes of this scale), that should be an occasion neither for shame nor for concealment.

The other cognitive phase, deriving from traditional scientific methodology, is called Data-Input. We use this name rather than "empirical", to include certain inputs (quite common in policy-related research), whose relation to controlled experience may be tenuous or even nonexistent. Starting again with the classical and strongest mode, we have *Experimental Data*. Not so strong, our next entry is *Historic/Field Data;* data of this sort is "accidental" in the sense of being taken as it occurs, and lacking tight controls in its production and/or strict reproducibility. Historic Data is that which was accumulated in the past, out of the control of the research process under consideration; Field Data is produced by procedures of collection and analysis, where the data are not subject to experimental control.

Historic/Field Data has at least the strength of a relatively straightforward structure, so that its possible errors and deficiencies can be identified. But sometimes data inputs are derived from a great variety of empirical sources, and are processed and synthesised by different means, not all standardised or reproducible. The numbers are then themselves "hypothetical", depending on untested assumptions and procedures. Even to estimate the spread and assessment in such cases may be quite difficult. Hence we assign *Calculated Data* to a weaker point in the scale even than the Historic/Field Data mode.

Traditionally, the last mode discussed above would have been considered the weakest in a scientific study. But with the emergence of policy problems calling for data inputs regardless of their empirical strength, formalised techniques were created whereby opinion could be disciplined so as to provide a reasonable facsimile of the facts. Such were subjective probabilities, Bayesian statistics, and other ways of eliciting quantitative estimates from experts. These we call *Educated Guesses*. Sometimes even such a mode is absent; guesses can be simply *uneducated* and yet accepted as data, hypotheses or even facts, whichever seems plausible. In this respect Data inputs in modern times have come a long way from the relative certainties of the classical methodological framework for science.

   The social aspects of the pedigree are here given in two phases: *Peer-Accep-tance* relates to the particular information under evaluation; and *Colleague Consensus* describes that aspect of the field in relation to the particular problem area. These are the phases to which users (and those who advise them) could turn first, for preliminary evaluations of the possible effectiveness of the information. Thus if there is weak Colleague Consensus and a research field is seriously divided (with *Competing Schools* or perhaps only *Embryonic*) then there will be no security in any piece of quantitative information. Even the sampling of expert opinions, to obtain Educated Guesses, can lead to a bimodal distribution or worse (Funtowicz and Ravetz, 1984); in this situation the policy maker learns the important lesson that scientific ignorance still dominates the problem. Stronger Colleague Consensus as with *All but rebels* or *All but cranks* may well be time bound. Since, as T. H. Huxley said:

It is the customary fate of new theories to begin as heresies and to end as superstitions (Mackay, 1977),

who is a "crank" or even a "rebel" depends on circumstances. There is a real distinction between the two cases; rebels have some standing among their colleagues, whereas cranks have none. At the other extreme from scientific orthodoxy, we have the mode *No Opinion*, where there is simply no cognitive framework or social network in which the profferred information can make any sense when it appears. This may be from the lack of apparent substance, or of interest, or both.

   Once we have an appreciation of the context in which peers can recieve and evaluate a piece of information, it is useful to characterize that process. The modes of Peer-Acceptance range in linear order from *Total* to *None*.It is important to realise that the significance of any given mode of Peer-Accep-tance depends critically on the state of Colleague Consensus. Thus if there is a strong general colleague consensus and weak acceptance the information must be judged as of low quality (assuming our trust in the general competence of the field). But if colleague consensus is as weak as peer-acceptance, such an adverse judgement is not appropriate. For example, if there are Competing Schools, every item of information will tend to be rejected by members of the opposing schools. There is a sort of converse case; if in spite of low Colleague Consensus an item of information achieves high Peer-Acceptance, then we know that it is of exceptional quality. We have split the "social" aspects of pedigree into two phases, in order to accommodate such cases as these.

   This third phase, Peer-Acceptance, differs in an important respect from the other three. Those may be said to characterize a whole area of practice, whose community is composed of those who are technically competent for quality evaluation of each other's work. (Clearly, one researcher may belong to a plurality of such communities). For such a community, the fourth phase must have one unique mode; one collectivity cannot simultaneously be both All but Rebels and Competing Schools. Although it is less obvious, there will tend to be a single set of modes on the cognitive side as well. (We will discuss the exceptions to this tendency later in this chapter). By contrast, the mode of Peer-Acceptance will depend strongly on the particular item of information in question. For this represents the communal judgement of quality; and therefore this is the only phase whose modes vary widely over the individual items of information in a field as we have defined it.

With these distinctions accomplished, we can be more precise about the way in which pedigree reflects the deepest sort of uncertainty, border with ignorance. The two cognitive phases, together with Colleague Consensus, define the state-of-the art of the relevant field; and in that sense they locate the common border with ignorance. In the case of an individual quantitative result, Peer-Acceptance will also reflect the state-of-the-art when its mode is the maximum for results in the field. If then contributes to the mapping of the border with ignorance. If the mode is lower than the maximum, then the item of information in question has been judged as of lesser quality; it lies within the area of attained knowledge, and does not contribute to the advance of the field beyond its existing frontiers. Thus the border with ignorance is defined generally by modes in the cognitive phases and in Colleague Consensus, and is completed by the maximum mode in Peer-Acceptance.

We may briefly review the modes and phases of this pedigree matrix, to illustrate the normative ordering. For example, if we qualify the theoretical structure of the production process of the information as a computational model (e.g. weather forecasts), we are implicitly stating that we do not have a theory-based model (as a hydrodynamical system), and so we record the absence of an effective theory. Similarly, if the data input is not experimental (as in traditional laboratory sciences), it can be at best historic/field data (as in environmental research). In the latter, data are inherently less capable of control, and so are less effective as an input and check for theoretical structures. On the social side, we can see the development of colleague consensus in the progression of revolutionary scientific theories. Thus for Einstein's theory of relativity, the modes moved from embryonic in 1905, through to all but rebels in the 1920's, and finally to all but cranks by 1950's.

10.5. APPLICATIONS OF PEDIGREE

Now that we have available the full machinery of NUSAP, we can revisit some of the examples discussed previously. We will use the pedigree codes and abbreviation of the names of the modes shown in Table V. There is an illuminating contrast between the examples of the fine-structure constant and the "pasture-cow-milk" pathway model for radiation. The fine-structure constant has a nearly-top pedigree (4,4,3,4). The 3 indicates high rather than total peer-acceptance, to distinguish the provisional acceptance of a recommended value which is certain to change. If we took a later date for the value, when it had settled-down, then we could increase this rating to 4. Of course, this total acceptance would not prevent a later change, as in the recent case of the gravitational constant. A full NUSAP representation is

TABLE V
Abbreviated research-pedigree matrix

Code	Theoretical structures	Data-input	Peer-acceptance	Colleague consensus
4	Th	Exp	Tot	All
3	Th.bM	H/F	Hi	All–
2	Mod	Calc	Med	Sch
1	St	Ed.G	Lo	Emb
0	Def	Gues	Non	No–O

Notation:

$$\alpha^{-1} \text{ (year)} = 137.0 + x : \text{E-4} : \pm \sigma_{ppm} : \pm 2\sigma_{previous} : (a,b,c)$$

Instance:

$$\alpha^{-1} (1968) = 137{,}0 + 360 : \text{E-4} : \pm 1 : \pm 2.6 : (4,4,3,4)$$

By contrast, the radiobiological model has a weak pedigree. The instance is

$$2 : \text{E-8 Sv} : \Lambda 5 : 50/1 : (2,1,3,1)$$

The state-of-the-art here does not extend to theoretically based models; many of the crucial parameters are obtained mainly by guesswork; within the field there is strong acceptance of this result; but the field itself is in no better than an embryonic state of development.

Now we review the prediction of increase in global mean temperature. The pedigree here is (2,2,3,1). How do we justify this coding? We can elicit it by analyzing the World Resources Institute's Model of Warming Commitment. Are there effective Theoretically-based Models for atmospheric CO_2 and its temperature effects? According to *Nature,* not quite; severe uncertainties exist on the time-scales both of millions of years and of days. Hence we have at best, Computational Models. What about the data that are injected into the models as inputs? There is some which are better than Educated Guesses but (as yet) not much obtained through instrumental readings, even as Historic/Fields Data. Hence we do best to call them Calculated Data, usually resting indirectly on measurements. Moving now to the social aspects of the pedigree, is the state of development of this field, as expressed by Colleague Consensus better than Embryonic? To have Competing Schools would require the presence of some realistic theoretically-based models; and these are not yet with us. In such a situation, Peer-Acceptance of the result is not such a critical indicator of its quality; here it seems high, giving the numerical coding 3. The full NUSAP expression for the forecast is

$$(1\tfrac{1}{2},\tfrac{1}{2}) : {}^{\circ}C_{2030} : - : \text{high}_c : (2,2,3,1)$$

We notice that the assessment rating is stronger than would be justified by the pedigree alone. We recall that this reflects the likelihood that the forecast includes the true value; and this is more a consequence of the breadth of the inexactness interval than either its scientific strength, or usefulness for policy. A NUSAP expression reflecting these other considerations would, as we have seen, have different entries is assessment, reading

$$(1\tfrac{1}{2},4\tfrac{1}{2}) : {}^{\circ}C_{2030} : - : low_p : (2,2,3,1)$$

We may now discuss various cases of quantitative information that were important in the development of science, and which illustrate significant features of our pedigree category. Not all quantitative information is appreciated on its first publication; the classic example is Mendel's simple arithmetic ratios between frequencies of different sorts of hybrid peas. For the first thirty years after its publication, the implied pedigree was, as seen retrospectively by historians:

(Th.bM,H/F,Non,No–O) or (3,3,0,0).

The zeroes reflect the nearly complete contemporary ignorance of Mendel's work. Of course, any contemporary who might have scanned Mendel's paper would not have been so complimentary on the cognitive side. A (reconstructed) pedigree code for Mendel's results as seen by colleagues would be

(St,Calc,Non,No–O) or (1,2,0,0).

The Calculated mode (rather than Historic/Field) conveys the suspicion that the simple ratios were the result of a coincidence or perhaps of "massaged" data. In the earlier twentieth century, the rediscovery of Mendel changed the pedigree to

(Th.bM,H/F,Tot,All) or (3,3,4,4)

With the further developments of genetics, the ratios themselves are strengthened to have a pedigree

(Th,Exp,Tot,All) or (4,4,4,4).

But greater sophistication in statistics and its application to experimental design, led to a scrutiny of the aggregated numbers by R. A. Fisher. He found them "too good to be true", and so the modern historians' judgement of Mendel's own work (Olby, 1966) in his own time now has pedigree

(Th.bM,Calc,Non,No–O) or (3,2,0,0).

That is, we accept that there was a Theoretically-based model, but we are not quite so sure about the data.

An example of an inverse development was provided by Kuhn (1961) in his seminal essay on measurement in science. This concerned the experimental value for a constant of crucial importance in the caloric theory of gases: the ratio of the two sorts of specific heat (at a constant volume and at a constant pressure). The setting for the production of this number was quite dramatic: the Laplacian theory of gases could (unlike any other) explain the experimentally determined velocity of sound in air, but only if the constant in question had a certain predicted value. The *Académie des Sciences* devoted its annual essay award competition to this topic in 1819; and the desired value was duly obtained by Delaroche and Berard, whose work won the prize. All was perfect; and here we have a pedigree

$$(Th,Exp,Tot,All-) \text{ or } (4,4,4,3),$$

the only reservation being among the nascent scientific/political opposition to the Laplace school. Unfortunately, the result was simply incorrect; and its background theory became discredited for many reasons. A retrospective pedigree for the result, a decade on, could be

$$(Th.bM,Calc,Non,All-) \text{ or } (3,2,0,3);$$

here the Colleague Consensus embraces the victorious anti-Laplacian party, comprising nearly all save the lonely disciple Poisson (Fox, 1974).

These two examples of the rise and fall of pedigree ratings for quantitative information provide a reminder that the evaluation of scientific results is a matter of judgement, which can change drastically. What is effectively scientific knowledge at any one time is very much liable to subsequent revision by the wisdom of hindsight. The pragmatic reliability of scientific information is not tightly bound to its status as correct knowledge. Thus, we still teach "calorimetry" and Newtonian mechanics in schools, along with many relics of obsolete and oversimply theories in chemistry. Such antiquated versions of scientific concepts remain alive by their use in all applications other than those at the research frontier. In this way they survive as "knowing-how" until such time as they are superseded by newer versions of concepts that have greater pragmatic effectiveness. The NUSAP categories of assessment and pedigree can be used in conjunction to chart the varied fortunes of items of scientific knowledge, in the complex interplay of its cognitive, social and pragmatic aspects.

10.6. A NUSAP EXPRESSION FOR A POLICY FORECAST*

Here we provide an example of the variety of ways in which uncertainty can be expressed through NUSAP, in the statement of the outcome of a study involving modelling with expert judgements as inputs (Funtowicz and Ravetz, 1986).

The study was conducted while the OPEC agreements were still in force; the task was to estimate what the price of oil would be on a competitive market. The outcome of the exercise was in the form of a probability function; the cumulative distribution in given in Figure 9. This can be interpreted as follows:

---

* This section reproduces research done with C. W. Hope.

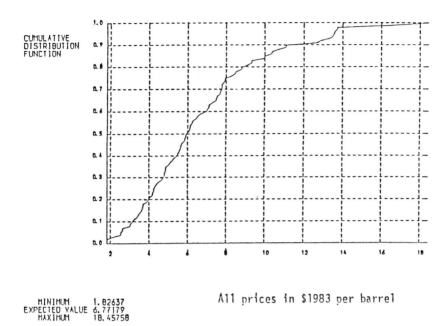

MINIMUM   1.82637
EXPECTED VALUE 6.77179
MAXIMUM   18.45758

All prices in $1983 per barrel

Fig. 9. The Competitive Price of Oil (Hope and Gaskell, 1985, 293).

In a perfectly competitive non-cartelized market, there is a 10% change that the price of oil would be below about $3 per barrel, a 50% chance of being less than $6 per barrel, a 90% chance of not exceeding about $11 per barrel. (Hope and Gaskell, 1985, 293.)

Translating this result into NUSAP is relatively straightforward. The centre of the probability distribution is around $6 per barrel, and in view of the spread of the values it would be inappropriate to try to put anything more precise (e.g. 6.02) in the numeral place. The price is measured in $(US) of the year 1983, expressed in the unit place as $_{83}$. The inexactness described in the original paragraph can be quite adequately represented as a factor of 2 either way, giving f2 or Λ4 as the entry in the spread position. It is clear that 80% of the probability distribution lies within this range around the central figure, so the entry for the assessment place is 80%.

Finally the pedigree of this result must be described. It was produced by a mathematical model working with a standard theory of competitive oil pricing. The data was not obtained by historical observation, but by a litera-ture survey of previous research plus some educated guesses, so it can be characterized as being Calculated Data. The theoretical structure upon which the model is based has never been empirically verified to anyone's satisfaction, but is has a long history (Hotelling, 1931) and is completely standard in research publications attempting the task. At worst, its use in this study could be expected to cause only partial disagreement. The Bayesian probability distributions of data input were constructed from an extensive literature survey, but they can only be claimed to have the status of the authors'

subjective opinions. It would therefore be prudent to put the level of agreement about data as fairly low. This combines with the general agreement over
the theory, to give a rating on the social components of pedigree around the
middle of the range. The NUSAP expression of this result reads,

$$6 : \$_{83} : f2 : 80\% : (3,2,2,2)$$

NUSAP can also be used to highlight the most significant result illustrated
by Figure 9, namely that the highest value for the competitive oil price lies well
below the actual oil price in 1983 of about $\$_{83}30$. The only difficulty in
converting the upper limit to NUSAP is to elucidate the meaning of "upper
limit". Clearly we should not take the highest value from the model simulations: if we performed 10,000 simulations we might expect at least one
unreasonably high value to be thrown up. Instead we associate the upper limit
with a significance level, 95%. Referring again to Figure 9, it can be seen that
95% of the model results lie below about $\$_{83}14$. The only spread on the value
comes from the sampling error associated with performing only a relatively
small number (100) of model simulations. Another 100 simulations could give
a value for this upper limit about $\pm 25\%$ different from $\$_{83}14$. The pedigree
remains unchanged since the data and model are the same as before. In
NUSAP then

$$14 : \$_{83} : \pm 25\% : \%95 : (3,2,2,2)$$

We notice the two notations used so far in the assessment category; '$N\%$"
described a result with an $N\%$ chance of containing the true value, "$\%N$"
described a specific N^{th} percentile of a probability distribution.

The numerous studies made since 1973 of the hypothetical competitive
price of crude oil constituted a body of evidence that was based on restrictive
theories, and which utilized historical data that commanded general agreement from experts in the field. These studies together qualify for a pedigree
(3,3,3,3), and their results represent the state-of-the-art, and therefore the
border with ignorance, as far as hypothetical competitive oil prices are concerned. It is fair to summarize the results of all these studies in terms of the
upper limit of the hypothetical competitive oil price. This is an order-of-magnitude, better represented by $10 rather than $1 or $100. This result can then
be expressed.

$$1 : \$_{83}10 : U_p : (3,3,3,3).$$

The unit entry indicates that conceptually nothing finer than a $10 graduation
is justified, while the hyphen in the spread place shows that no estimate of
precision would be meaningful. The entry U_p in assessment, for "upper", is
appropriate since experts would not express this information as a specific
percentile of a probability distribution.

The three NUSAP representations form an instructive contrast in their use
in the policy context. The first provides an estimate of the expected oil price
in the absence of the "cartel"; this has clear implications for many policy

decisions. We note that even to within a factor of 2 either way, calculated prices are far below their actual 1983 level. The second result makes this aspect of the communicated information explicit, by estimating an upper limit (with a rather narrow confidence band) which is still only a fraction of the 1985 price of crude oil; it provides an indication of the strength of the forces inhibiting a competitive market. The third result represents a rough consensus of people knowledgeable about the market, which, gives a reliable answer to the policy question addressed by the study: Could the actual oil price be the outcome of actions by competitive producers?

The existing standard notations, as with significant digits, are not powerful enough to convey such differences in meaning, which are so important when the quantitative information is to be used as an input to a policy process. In particular those notations give none of the protection against spurious claims of precision and consensus, that the assessment and pedigree categories of NUSAP automatically provide.

10.7. PEDIGREES FOR ENVIRONMENTAL MODELS AND FOR FUNCTIONAL QUALITY

In policy processes for decisions on environmental problems, the scientific inputs are frequently obtained from computer models. We have already seen how controversial these can be. In the debate we have described (Chapter 1), it was clear that in the absence of the classical procedures of scientific validation, the guiding principle in their evaluation is quality assurance.

The pedigree matrix for such models is given as Table VI. We see how the cognitive phases run parallel to those of the research pedigree matrix. The modes given here for model structure are based on the analysis of Beck (Chapter 2). For Data-Input, we notice significant modifications of the classical epistemological ordering. First, since the data are not generally produced in the project in question, Review is the best mode. This reflects the need for prior selection and evaluation of available data entries. Then we notice that

TABLE VI
Pedigree matrix for environmental models

Code	Model structure	Data-input	Testing
4	Comprehensive	Review	Corroboration
3	Finite-element approximation	Historic/ field	Comparison
2	Transfer function	Experimental	Uncertainty analysis
1	Statistical processing	Calculated	Sensitivity analysis
0	Definitions	Expert guess	None

Experimental is placed lower than Historic/field. This is because laboratory data will generally not reflect the field conditions to which the model refers. (We will discuss this in more detail in connection with radiological models in Chapter 12.) For the phase of testing, the modes relate to the different sorts of procedures for the validation of models. The best is *Corroboration*, meaning some sort of testing against independent field data. Failing that, we may have *Comparison* among models that to some degree are independent. On a single model, we may conduct an *Uncertainty Analysis*, which examines the different sources of uncertainty. At the very least can be a *Sensitivity Analysis,* in which parameters are varied through a range of values, and the variations in the calculated model output observed.

When we move from the cognitive to the functional aspects of information, the task of assessing quality becomes much more complex. This would not be so if we were dealing with traditional scientific information, destined to be used by a well-defined community of colleagues, sharing the same family of techniques and also the same goals. In the extreme case of pure science, there is no effective difference between cognitive and functional quality, since what makes a result excellent is its combination of internal strength with a promise for development of more results similar to itself. Here, in the area of policy-related research, it is quite otherwise. As we have seen, there are many sorts of actors, each legitimately possessing their own agendas, valuations and perceptions. The great lesson that scientists of this generation are learning is that these actors will not defer to the research scientist's special vision of problems and solutions. For better or for worse, they will demand a hearing for their own. Accordingly, their management of uncertainty, and their criteria for quality, will be different from those of the scientist, and indeed will differ among themselves.

The classification of "Critical Criteria" by Clark and Majone (Table 3, p. 61) organizes all this variety into a table which even in its original version nearly has the form of a pedigree matrix. We have a list of actors, from Scientist over to Public Interest Group; these define the rows in the table. Defining the columns are the three "critical modes", those of Input, Output and Process. The boxes in the matrix, corresponding to a particular critical mode for a particular actor, contain a set of criteria, usually four but as little as three or as many as six; some are expressed as phrases, others as questions.

We may consider this table as a family of partial pedigree matrices, applicable to any of the actors defined there. For any given actor, we can arrange the "critical criteria" in each box in the forms of modes under the three phases. We lose the rectangular shape for a pedigree matrix, but we gain the richness of the analysis of these authors. Since in this case the modes are independent, there is no question of a normative ordering; and the numerical codification for any given result must be accomplished by a special arithmetic.

Given an actor on whose behalf we are analysing functional quality, we first see which of the three phases (Input, Output, Process) are of concern to them. Those which are not, are crossed out immediately, and do not enter into the scoring. Then in each remaining phase, we examine the modes in the same way; again we eliminate those which are not relevant to our actor's concern. Of the modes that are left, we complete the checklist with a simple scoring, for example 0, $1/2$, 1, representing quality for each criterion. These are summed for each phase, compared to the maximum possible, and then normalized to a full score of four (to bring it into harmony with the other phases of the pedigree code). We can refine this procedure by scoring 0, $1/2$, 1 for relevance as well, and then summing over the product of Relevance and Quality. All this is quite rough and ready, but it enables the translation of the "critical criteria" table into an useful gauge. On its basis, an assessment rating for pragmatic or functional quality can be made when it is desired.

As an example, we take again the prediction of global warming. From the point of view of policy making in general, it is only the Output that matters; so we neglect the other phases. Our checklist is now very simple:

	Relevance	Quality	$x \sim 4$
Is output familar and intelligible?	1	1	1
Did study generate new ideas?	$\frac{1}{2}$	0	0
Are policy indicators conclusive?	1	0	0
Totals:	$2\frac{1}{2}$	1	1.6

The resulting score, which may be included in the full pedigree code or taken over directly into Assessment, is between 1 and 2. Had the study been expected to generate new ideas, the criteria would have been slightly more demanding with a Relevance total of 3; and the normalized score would have been 1.3. We notice that when there are less than four relevant critical criteria, an item which fully satisfies a single one takes the score beyond the minimum rating for content. In this case, we have no reason to round up the 1.6 to 2; a rating of 1, which fits our prior evaluation, is appropriate.

Thus the Clark-Majone classification provides us with the relevant phases, one to three in number, to adjoin to any other pedigree matrix when an evaluation of functional quality is desired.

10.8. ELICITATION: USE AND DESIGN OF PEDIGREE

We have provided an introductory example of elicitation, in connection with spread (Chapter 9). We recall that this went in two phases, the first establishing the topology and the second the estimate within it. This elicitation included questions that could appear surprising or even simplistic, so as to help the subjects get away from their unselfconsciously accepted frameworks. Even to achieve clarity about the numeral and its appropriate expression may turn out to require an enhanced awareness of the production process for that number.

Our examples of some important numbers for science and policy, such as the fine-structure constant and the temperature rise forecast, show how every such reported number derives from a production process which itself has a history. The purpose of the pedigree matrix as we have shaped it, is twofold. First, it provides a framework for analysis of that production process, and then it can also serve to guide the conduct of the analysis in any particular case. Most simply, it functions as a checklist for the elicitation procedure.

Elicitation of the pedigree codes for a particular quantitative result can be a very simple procedure. Like other elicitations, it works best if the experts are completely familiar with their tools and also have a good acquaintance with the relevant field. With such craft skills, they can avoid the pitfalls of the work, which are mainly in the personal alienation of the subject from the elicitation procedure, so that inaccurate or misleading impressions are obtained. Familiarizing the subject with the NUSAP approach is, of course, essential. Although the identification of four relevant modes for an item of quantitative information does not take much time in itself, the exercise should not be like a multiple-choice exam. This would be a very narrow interpretation of an elicitation, producing minimal understanding.

It should not be surprising that in this, the most qualitative of the NUSAP categories, effective work requires appropriate craft skills based on judgements. These appear in several contexts. First, it is not always a straightforward operation, to decide which mode applies to the field in which a particular quantitative result is produced. In some cases (and this is particularly marked in policy-related research) a single number may be the outcome of processes that are described by several modes. For example, it may occur that the separate parameters in a model derive from data which is experimental, historical/field, calculated and even guessed. How to code for such diversity? One way is to see which data are most critical for the quality of the results; put otherwise, which are most likely to vitiate it, should they be unreliable.

If there is such a single dominant source, then that can be used to define a focus for the gauging of the appropriate representative mode for that phase. If there is a range of modes that cannot be ignored, then notational devices, such as double entries (e.g. 2/3) or even intervals (e.g. 1–3 or $j = 1,...,3$) may be used (Macgill and Funtowicz, 1988, 79). Fractions, expressing "average" of modes, are in the strict sense meaningless, and are to be avoided at this stage. The guiding principle in representation is that hyper-precision is to be avoided; a rough coding which is easily understood is better than one with refined distinctions, giving satisfaction to the author's conscience but to little else.

A similar problem arises when full NUSAP expressions, including pedigrees, have been obtained for the different elements of a simple sequential model. Which of these is to be accepted as the pedigree code for the final result? We will later discuss a technique whereby the propagation of some uncertainties, as spread and assessment, may be mapped from the separate parameters through the calculated model output. There is no analoguous algorithm for pedigree, especially since different sorts of pedigree matrices may apply for the different parameters. The solution of this problem is achieved by considering pedigree, not as an abstract exercise, but as a tool for the evaluation of useful information. The question of the pedigree coding of a quantitative model output becomes relevant when that number is to be used as an input to another process, perhaps scientific, perhaps policy. Then it is an item of data among several or many; and it needs to be evaluated as such. We will develop and discuss an appropriate pedigree matrix for this function in Chapter 12.

The basic design criteria for NUSAP should also guide our use and evaluation of a simple system for using pedigree ratings as a guide to assessment. We have mentioned the possibility of using the numerical pedigree codes as a sort of gauge, and combining them in an artefactual arithmetic to achieve a score, or grade, as a notation for the assessment category. The procedure is as follows: with codes given on a scale from 0 to 4, their sum can range from 0 to 16; and that interval is then divided into ranges with quality descriptions or grades. We shall provide an example of this technique for a Reliability Index, in connection with statistical information in the next chapter. The very simplicity of this technique may lead some users astray; for it is only a rough guide, to be used when there is no other information available for assessment. A warning signal is the case when the modes of the different phases of the pedigree matrix are quite different in their ratings. This may indicate that there is something not quite standard about the history of the item of information; and a simple sum of the scores can obscure important aspects of its production and quality.

A very important element of the skill of applying and interpreting pedigrees is an awareness of what sort of judgements and evaluations they express. Pedigree has been designed to be intersubjective; this brings strengths to NUSAP, but also imposes limits. Our experience has been that there is a remarkable degree of agreement among experts on the pedigree ratings of results within their general competence. Even when there are sharply differing evaluations of the work of colleagues in competing schools or neighbouring areas of research, it seems that the state-of-the-art of the different fields, as broadly defined by the phases and modes of the research-pedigree matrix, are uncontroversial. Thus while the pedigree ratings cannot in any way aspire to the status of objective science, they do seem to be sufficiently robust to qualify as genuinely intersubjective. For the distinctions between the separate modes correspond to very basic understandings among scientists; once the definitions of the modes are provided, there is very little room for disagreement. Only in the cases where the sources of information are very mixed will there be differences over the best form of representation.

The intersubjectivity of the pedigree category is a great source of strength for the whole NUSAP system and approach. Its skilled use will increase consensus, and also enhance awareness, on uncertainty and quality of information, without embroiling its users in debates about peripheral issues. If there are queries about the quality of scientific information in a policy issue, NUSAP can serve to establish some explicit common ground, for an orderly dialogue. Our stress on intersubjectivity serves as a remainder of the complementarity between the more and the less formalized aspects of evaluation. As we have seen, NUSAP is designed to guide judgements and improve skills, and not to replace them. The dream of a comprehensive, universal pedigree, in which all possible evaluations can be coded, is alien to the NUSAP approach.

Such principles underly our approach to the evaluation of the craftmanship which has gone into the production of particular quantitative results. The Peer-Acceptance phase of the pedigree reflects evaluations of this, most noticeably when the mode recorded is not at its maximum possible level. This phase expresses an evaluation in general communal terms, rather than a fully articulated evaluation by an expert with personal knowledge of the researcher and the work. There could indeed be a pedigree matrix for guiding an elicitation of this aspect of information. But it would need to be designed on the basis of close acquaintance with the circumstances, conditions and materials of research in the particular field; otherwise it could all too easily be either too general, or too impossibly detailed, to be of real use.

In the simpler cases of extension of existing pedigrees, it may occur that only a refinement of modes is necessary. We may imagine the process as opening "windows" for the relevant modes; these may be of a general or special character. In this way, our two basic pedigree matrices (the Statistical Information pedigree will be developed in Chapter 11) may be used as frameworks in which more specialized pedigree matrices may be inserted. These two express the most fundamental aspects of the production process of information: the complementarity of cognitive and social phases and the hierarchical ordering of modes.

In designing new pedigree matrices, our first and basic criterion of simplicity must be adhered to in spite of many temptations. The evaluative judgements which are involved in pedigree may become very complex and interdependent, particularly when one descends from the very general level of analysis exhibited so far here. We have quite deliberately kept our matrices in rectangular form, employing the same number of modes in each phase. This is not only simpler to visualize; as we shall see it also enables the use of a simple artefactual arithmetic for the construction of scores and grades to be used for the representation of assessment. Doubtless a careful analyses would be able to yield sets of modes with finer distinctions, and of varying length, for each of the phases. But the gain in analytical precision would be achieved at a heavy cost in simplicity and effectiveness. Other more complicated devices for conveying nuances of meaning within pedigree will also suggest themselves. In every case we must keep in mind the design criteria for NUSAP and the conceptions on which it is based. A great strength of NUSAP is its ability to detect hyper-precision of expression and indeed to train and sensitize users in the skills of appropriate precision in quantitative statements. The skill of developing new NUSAP notations is at a higher level than that of using them, and the pitfalls of hyper-precision are correspondingly more dangerous.

We have seen previously how the pedigree ratings can help to chronicle the rise and fall of scientific theories and fields. But they cannot provide a basis for an Olympian judgement on scientificity. When there is a consensus on the character of a field, past or present, a pedigree elicitation can express that consensus in an illuminating way. Where there is no agreement, the pedigree elicitation procedure can not by itself settle the issue. Thus, for fields that are obsolete (e.g., ether-theory) or out of fashion (e.g., descriptive geometry) there can be consensus. But, particularly in policy-relevant areas, there can occur fields whose status is violently contested. In such cases, there may be a tightly-knit group of insiders, who reject all critics as rebels or even as cranks. Then for any important item of information, there will be alternative pedigree ratings, which may differ in nearly all particulars. The NUSAP approach cannot resolve such fundamental problems in a straightforward way. However, experience shows that there is always an ongoing dialogue about policy-relevant fields, involving the experts, with their patrons and intended users, as well as critics both inside and outside. Fashions change in these areas even more rapidly than in academic science. In this fluid situation, where knowledge and ignorance are so intertwined that traditional categories scarcely apply, the intellectual discipline and practical guidance provided by NUSAP can still help for the clarification of debate.

CHAPTER 11

THE NUSAP PEDIGREE FOR STATISTICAL INFORMATION

Our first exercises in developing NUSAP were in connection with information used for research. Typically this involves a small set of data-items, usually the results of experiments but sometimes (as in policy-related research) based on personal judgements and estimates. This information would be created in the course of the investigation of some particular problem, and would take its shape from that context. We now apply the NUSAP system to another kind of technical information. This is "statistical", in the sense of having to do with statecraft (Funtowicz and Ravetz, 1989). In modern times the power of quantitative statistical information was first demonstrated by William Petty's *Political Arithmetic* in the seventeenth century. Since then, census-taking has become a responsibility of all governments, and has become an ever more elaborate and sophisticated exercise. In the case of the United States of America, it is related to the constitutional duty of the President to give an annual report to the people on "The State of the Nation".

There have always been two conceptions of the sort of knowledge that is obtained through statistics. Some have believed that this is a fully-fledged quantitative science, the "physics of society", producing facts from which policy conclusions flow by logical entailment. Others have been more pragmatic, accepting that statistics can provide only policy forecasts and not scientific predictions. Thus when insurance companies lose money in some field, they do not speak of a scientific refutations of their actuarial techniques, they simply raise premiums. Here we will not try to adjudicate between these two conceptions, which relate back to the distinction between "knowing-that" and "knowing-how". For us it is sufficient to clarify the issues of uncertainty and quality as they affect statistics. For this we will give special attention to "indicators", as the means by which statistics accomplishes the connection between policy and empirical realities.

11.1. STATISTICAL INFORMATION: ITS PRODUCTION IN BUREAUCRACIES

For the proper understanding of statistical information, and also for its most effective utilization, we must appreciate its important differences from the "research" case. We are here analyzing the process where data are not collected for the investigation of some particular research problem, but as part of a store of information available for possible future uses in a policy context. Their significance may be less in their absolute quantities than as elements for

156

comparison in a time-series. Normally they are collected in widespread field operations, and are then subjected to a sequence of operations of aggregation and interpretation. The "theoretical" component constitutes the framework in which the field-data are categorised, rather than furnishing hypotheses to be tested as is typical in the "research" case. All these properties make it necessary for the work to be done institutionally. We note that more than one institution can be involved in the process; and independent individuals can "borrow" data provided by the original institution, for review and publication (i.e., "secondary analysis"). Such "external" agents can be seen, for the purposes of the pedigree analysis, as simply another element of the sequence of operations whereby field-data is finally converted into statistical information.

When we consider the production of each piece of information in organizations, several features are relevant:
- The context is of some "mission" that defines the organization, gives it a reason for existence, and shapes its work and institutional culture.
- The work is done in such a way as to ensure consistency and continuity within the institution, and effectiveness for its outputs.
- There is an explicit division of labour in all processes, and explicit definitions of individual tasks.
- Every task as defined is tightly embedded in structures of instruction and of evaluation.
- The methods of work, and even the concepts being manipulated, are defined explicitly through superiors in the organizational structure.
- The task of evaluation, accomplished informally in the "research" case by the scientific colleague community, is done formally here through review procedures, either internal or external.

The foregoing features define an ideal whose correspondence to reality is not automatic. Hence we add others:

- A complementary set of unofficial networks and practices exists, which may modify or even nullify those of the official part.
- There is no limit to the degree to which an institution can temporarily drift towards isolation from its external environment, whether it be the public to which it is nominally answerable, or the ostensible object of its officially assigned tasks.

In such circumstances, the main function of statistical information may become internal, in strategies for the avoidance and manipulation of uncertainty. In the absence of collegial, informal social mechanisms for the assurance of quality, there is no lower bound to the quality of bureaucratically-produced statistical information. (This will be reflected in our pedigree modes "symbolic" and "fiat").

This last tendency is well recognized, and external audits are a regular feature of any properly run bureaucracy. These will normally derive from a higher authority within the same system. Such audits tend to concentrate on procedure, perhaps extending to performances, and only rarely on policy. Although policy is constitutionally within the province of the formal political systems, there is now an increasing tendency for political debate to focus (perhaps unfairly) in the inadequacies of the bureaucracies in performing their stated functions. This occurs when the bureaucracy is allowed, or even pressured, to drift from its stated objectives and indeed from reality. This can be accomplished by the political or executive sectors crippling their bureaucracies through restrictions on resources and enforcement powers. To survive at all, the regulatory bureaucracies then tend to devote all their efforts to the avoidance of uncertainty and, should that fail, to its manipulation (we discussed this phenomenon in Chapter 2). In such circumstances, genuine "external review" is performed outside the institutional framework, by the criticism of NIMBY and pressure groups. These are now increasingly accepted as a legitimate part of the constitutional process in a democratic society.

All these considerations apply to policy-related research when conducted within bureaucracies. In the absence of regular, effective criticism and quality-control, there will be a "Gresham's law" of quality of information or research, the bad driving out the good. This is most obvious in the case of military R & D, as epitomized in the Strategic Defense Initiative or "Stars Wars" project.

Of course, none of these features are unique to bureaucracies; even in traditional laboratory science, there is something of an hierarchical division of labour, and the "bootlegging" of research and production of vacuous results are not unknown. But traditional science has operated in a style involving remarkably few formal structures; and philosophers and sociologists of science have generally neglected bureaucracy in the production of knowledge. Also, the really interesting science, traditionally, is that which has been successful; the phenomenon of an empirical scientific field having little contact with reality has not generally been considered worth analyzing by historians. Hence the phenomena specific to the field of institutionalized knowledge-production are still to be explored.

In the following chapters, we will use the NUSAP system to analyze cases of policy-related research, and show how we can begin to achieve quality assurance there. For the moment, we shall concentrate on organizational structures at the opposite pole from traditional science. These are involved in the sorts of statistical information with which we are concerned here. Complaints and uncomplimentary jokes about bureaucracies and their workings are doubtless as old as the institution itself. But we cannot assess the problems of this mode of operation unless we first appreciate its necessity. For this, let us take a very simple example of the definition of an object studied in the policy context. Suppose that someone wants to know the "total seating capacity" of the auditoriums in a city, perhaps for promoting the city as a conference centre. The number reported for "total seating capacity" is not a scientific fact, but is a policy indicator, constructed for the evaluation of the quality of the city in this respect. A survey is done; but in the absence of uniform definitions, many kinds of things are included. Some enumerators simply count the chairs they find in each hall when in use; others estimate from the floor area, allowing space for aisles; still others include standing-room and estimate the sitting-space on steps; others add the floor-space on stages where particular audiences (children, or Orientals) may be expected to sit or squat; some accept the local Fire Department's posted limits, and others do not; some even argue about the definition of "auditorium", on a continuous scale from an opera-house to a Scout-hut.

If each surveyor simply counts what is obvious to his or her individual common sense, an incoherent and useless number will emerge from the aggregation. To achieve an aggregation of numbers representing roughly the same thing, there must be prior decisions on definitions (related to the possible anticipated uses of the statistics); routines for ensuring uniformity among those in the field; and quality-checks of various sorts at every phase. If any aspect of this work of constituting and reviewing the inquiry becomes sensitive or contentious to any participant or observer, then standardized procedures for debate and decision must be available for use. There will also be the inevitable set of informal structures and practices, which must be appreciated by those whose use of the information will be crucially affected by its particular concealed biases and distortions.

Simply to have the same labels for data is no guarantee that the data will all refer to the same things. This is most clearly seen when data collections are separated by space, time or institutional location. When the indicators being quantified are of a more abstract character (as "seating" compared to "seats"), pitfalls in interpretation are more likely. Because of this, both comparison and aggregation of data sets require an elucidation of the operational meaning of their defining terms. In the above example, we could imagine either several neighbouring cities adding their "auditorium seating" data in spite of divergent standards: or a growth-rate calculated for one city over some years, with a changing data-base. Such variations may reflect differences in what is a "reasonable" or "appropriate" measure, in terms of existing practices or policy functions. For example, "seating" will depend very strongly on the state of local concern with safety. To aggregate or compare data gathered under different circumstances of safety-consciousness could yield very misleading results.

11.2. THE PEDIGREE MATRIX

We recall that the pedigree category in NUSAP is not only an evaluative account; it also has the function of expressing ignorance through a mapping of its border. In the case of "research" information there is an obvious reference to an epistemological taxonomy in the pedigree; the names of the various modes and the scalings relate to traditional understandings of the strength of scientific information. In the case of statistical information, the philosophical problems are as yet little explored by scholars; hence the epistemological aspect of this pedigree matrix may be less obvious. But when we reflect on how institutions accomplish the social construction of knowledge, we see that the border between knowledge and ignorance is complex, and that its mapping will require well articulated analytical tools. By their means we can describe how an institution can either achieve usable knowledge or create official ignorance, depending on its operations and internal culture.

Every entry in a table of statistical information has a complex history of its own; it is the product of both formal, reproducible operations and of personal judgements at various points. To provide a full historical account of each such entry, as an explanation of the table for potential users, would be an impossible task. The brief fine-print annotations on such statistical tables or the appendices at the back of the book, must normally suffice. The pedigree given here is designed to provide an evaluative account of the relevant aspects of the statistics, in a convenient codified form. It cannot convey the full account; nor is it designed for the explicit communication of judgements of crafmanship or of the pragmatic effectiveness of the technical work. Rather, it expresses the salient modes of operation in the different phases of the production of information. The phases are: *Definitions & Standards, Data-Collection & Analysis, Institutional Culture* and *Review*. The pedigree matrix for statistical information is displayed in Table VII.

By the first phase of this pedigree matrix, Definitions and Standards, we understand: all those decisions, logically prior to actual enumeration and testing in the field, concerning the establishment of the relevant conceptual objects and the set of operating procedures. We have seen from the previous example of "seating" that if data-collection is guided merely by common sense, then a totally incoherent set of numbers may emerge. This is of course a matter of degree: "chairs fixed in a position" is a fairly straightforward object for counting; "seats" is rather more abstract and therefore elastic; and "seating space" is even more so. As we become more abstract in our concepts of what is to be counted, we need more clarity in what is to be considered as a candidate for counting.

TABLE VII
Statistical information pedigree matrix

Definitions & standards	Data-collection & analysis	Institutional culture	Review
Negotiation	Task-force	Dialogue	External
Science	Direct suvey	Accommodation	Independent
Convenience	Indirect estimate	Obedience	Regular
Symbolism	Educated guess	Evasion	Occasional
Inertia	Fiat	No-contact	None
Unknown	Unknown	Unknown	Unknown

Some objects of data-collection are of a compound character; their constitutive parts may have all possible positions in the scale from the quite concrete, over to the abstract and the theoretically constructed. For example, in developing countries, major health and environmental hazards come from polluted drinking water. This is the object of the United Nations "Water Decade" of the 1980's (Agarwal *et al.*, 1981), devoted to the provision of safe drinking water and adequate sanitation for all the populations of the developing nations. Progress in the first part of this programme is officially measured by the indicator: "proportion of the population (rural or urban) with access to safe drinking water" (World Health Organization, 1984b, 154). Ordering these components from the simpler to the more complex, we have: "water", "proportion", "population", "rural", "drinking", "safe", "access". The first item is a natural object; the second a purely mathematical entity. "Population" can in principle be determined by well-known demographic techniques; while "rural" may require some conventions (as on maximum size of a settlement for it not to be counted as "urban"). "Drinking", of water, relates to use of particular supplies, and has its effective meaning determined by the last two components. "Safe" is in some respects a scientific category. But this judgement depends on a very complex process. It starts with scientific principles of bacteriology and toxicology, extends through standards as defined for laboratory practice, over the testing under field conditions, and also to quality-assurance on results of tests, and finally, includes judgements of "acceptable risk". In this case, the technical aspects of the term are intimately involved with institutional and cultural aspects. The final component of the indicator, "access", is as abstract as "safe", but lacks even its partial basis in scientific research. "Access" relates so strongly to habits and expectations, and even to the cultural meaning of the activities of obtaining water, that any simple conventional standard (as the original criterion of 100 meters from the household) is almost useless. Our rule for evaluating such compound indicators is generally by the weakest significant link. Thus, even if "rural population" is quite well defined in some place, and "safe drinking water" has some reasonable scientific meaning, "access", a component to which any aggregated statistics are highly sensitive, will, for us, determine its characterization for pedigree.

Under the heading Definitions and Standards we have five modes: *negotiation, science, convenience, symbolism* and *inertia*. The mode science refers to those cases where the definitions and standards are firmly based on scientific knowledge that is fully relevant to the situation and adequate in strength. The occurs rather less often than might be thought. The existing knowledge may need to be adapted to local circumstances, for which local experience is essential. Or the scientific inputs may not fully determine the policy conclusion; and then a negotiation with local interests is necessary and appropriate. Thus we have here an example where the traditional epistemological hierarchy is re-ordered; in such cases, negotiation may give better "knowledge" than science! Where the background science is inadequate for a full specification of the definitions and standards, several alternative modes are possible. We speak of convenience when those actually doing the work adapt the definitions and standards that are imposed from above to their particular circumstances. There are extreme cases where the scientific basis is nearly, or completely, irrelevant to the interests of those who ultimately control the institutional task. The need for institutional legitimacy or prestige, or other manifestations of values, may come to dominate over all other considerations; we speak then of symbolism as the mode in this phase. It can be detected by the presence of definitions and standards, which, however technical their form, are known to be wildly inappropriate or simply irrelevant to the task. There is a still weaker mode, inertia, to characterise the state where noone on the job now even knows or cares why the particular definitions and standards came to be used in the first place.

The order in which the modes are displayed is roughly normative; clearly science and negotiation are better than inertia. We put negotiation over science because it represents a richer contact with the real policy dimensions of the task. Symbolism is fairly far down, because it can produce results with very little real content. However, these rankings are rough; and the proper evaluation of the information is built up through all four of the loosely related phases of this pedigree.

The second phase, Data Collection and Analysis, refers to the operation in which field data is gathered and then analyzed and reported. Such work is inevitably a sequential operation; analysis comes after collection and will itself usually include several distinct levels. There are three sorts of possible practice in the measurement of policy indicators through field data. It may be impossible or unfeasible to measure directly (as, for example, total annual mileage of cars in some country); hence there can be an *indirect estimate* using some other variable related to the indicator by various intermediate assumptions. (We recall the example, total fuel consumption and a standardised mileage conversion factor, as in Mosteller, 1977). Of course it is sometimes possible to have a *direct survey*, as in a census, where various policy indicators are measured all at the same time. Every inquiry benefits from being completely designed around its own particular problem, with special care taken for characteristic pitfalls. Thus, paradoxically, a *task-force* dedicated to an intensive partial study of a problem, may yield better information about the phenomenon in all its complexity and variability, than a routine census survey which is nominally comprehensive. Coming to the other end of the scale, it may well be necessary to provide some information even when field data is simply unavailable; for that, an *educated guess*, based on expert opinions, may be appropriate. Finally, we have *fiat*, where some authority decrees the number which shall officially represent the phenomenon in question.

The subsequent phases of this pedigree refer explicitly to the organizational character of the exercise. Institutional Culture characterise relations between different elements in the sequential operation. In most institutions there is a formally defined hierarchy, where instructions go "down" and information goes "up". In the present case, definitions and standards come from "above" and field data from "below". Superficially, upper levels are related to lower through instruction and review; and the inferiors are expected to have *obedience* to their superiors. This is rarely seen in the pure case; and there are two contrasting sorts of modifications. Sometimes there is a two-way flow, *accommodation*, resulting in a *modus vivendi* whereby tasks are performed more or less to everyone's satisfaction. Where this is not possible, inferiors will resort to *evasion*, where only the formal aspects of the tasks are satisfied, to the detriment of their real content. The extreme case of this phenomenon is where orders from nominal superiors are simply ignored; we call this *no-contact*. There is another extreme, where the formal hierarchy is strongly modified and softened by shared interests and open communication; this we call *dialogue*. We give the best rating to this, analogous to negotiation in the first phase.

In the last phase of our pedigree matrix, we analyze Review. In traditional scientific practice, this is accomplished through largely informal procedures, as the peer-review of projects and the refereeing of papers. In market-sector enterprises, "control" is partly on quality of product, exercised through special units (and also by purchasers); and partly on process, through "audit" functions. Japanese industry has shown that high quality of work cannot be attained through simple obedience, but needs commitment and dialogue among all levels of an organization. This will be even more so when information, rather than material artefacts, is involved.

Every properly run organization will include arrangements for a *regular* review of its work. Normally this will be done by some special branch, operating routinely. For certain purposes, an *independent* review is used; typically this is done when unbiassed judgements are important for the assessment of some product or process. In special cases, the review might not be merely staffed by outsiders; it might be organized on an *external* basis; this would occur when the organization as a whole, or some particularly crucial or controversial aspect of its work, is subject to question. Coming down the scale, we have a situation where review, or audit, is at best *occasional*; or indeed where there is *none*. In the absence of review, there is ignorance on the quality of the products of the task, and quality assurance is nil. The evaluation of this phase should then make no positive contribution to the assessment rating of reliability.

This phase, Review, rounds off our picture of organizational operations as conveyed by this pedigree. It is complementary to Institutional Culture, as both relate to the institutional style; while Definitions and Standards and Data Collection and Analysis relate to the work. It is particularly important to include this last phase in a study of statistical information produced by bureaucracies, for of all organizational cultures they are the most prone to drifting out of contact with reality. An institution that lacks means of enforcing review and revision of its tasks and methods will inevitably become self-serving and isolated, and so will eventually lose its legitimacy.

11.3. PRACTICAL PROBLEMS

For the sake of simplicity we have assumed up to now that the data have been collected in the field, and then sequentially processed up through different levels of a single organization. In many important cases, users of information are external to the institutions where the data are collected. The pedigree matrix developed here also applies to that situation. External users of information provided by an organization can have a great variety of relationships to it. If they are in a superior institution, as a central government agency with subordinate though distinct regional agencies, then the original interpretation of phases and modes applies. The hierarchy may be weaker than this, as in the case of federal governments (such as the American, German or Canadian). It may be a strong transnational organization like the EEC, a weaker one like the United Nations, or an association like OECD. In these examples the relationships of authority are progressively attenuated. They are quite absent in the case of individual or corporate users of information operating outside the institutional nexus. Such persons might be experts solving particular problems in a symbiotic professional relationship with the institution; or they might simply be readers of published works where the information is contained. These last elements in the sequence of processing can handle the data in many ways. If they merely copy or report them passively, there is no change in the pedigree. On the other hand, they may manipulate them or even interact with providors at any previous stage of the sequence. These various possibilities should be reflected appropriately in the pedigree of the information they produce.

The problems of aggregation of incoherent data-sets are reflected in the pedigree matrix at several points. It frequently happens in practice that those who aggregate data find themselves with data-sets which are incomplete or incoherent in some way or other, and where they have no power to improve this material. They are inevitably forced to make policy decisions about the "best" way to render such imperfect materials in the aggregation process. Perhaps the simplest problem is when one constituent is so large compared to the others that its contribution, and eventually its changes, make all the others irrelevant. Whether one aggregates large and small constituents, depends on the context; if all that counts is the total, regardless of the distribution, then aggregation is justified. But users of the data should be made aware that the smaller constituents are swamped by the larger, and hence no effective information is being conveyed about them. Uncertainty concerning their statistics is drastically increased.

More complex problems arise when data-sets have inconsistent definitions or are incomplete in some way. Then aggregation and comparison becomes a skilled craft; those who do it must be quite sensitive to the costs and benefits of any particular procedure for possible classes of users. It is in this problem that we can see the close relationship between the different phases of the pedigree for statistical information. Any imperfections in Definitions and

TABLE VIII

Reliability index for assessment and abbreviated pedigree matrix for statistical information

	Assessment			Pedigree		
$\sum \geq$	Rel. index	Code	Def. & St.	D.C. & A	Inst. cult.	Review
13	High	4	Neg	Task	Dial	Ext
10	Good	3	Sci	Dir	Accom	Indp
7	Medium	2	Conv	Indir	Obed	Reg
4	Low	1	Symb	Guess	Evas	Occ
0	Poor	0	Inert	Fiat	No–c	None

Standards and lack of communication reflected in Institutional Culture, must be managed in Data Collection and Analysis; and finally scrutinized in the phase of Review.

In Table VIII we display the categories of the Statistical Information pedigree, with their abbreviated names. Also, the Reliability Index, derived by summing the numerical codes of the entries in the various modes, is illustrated. We should keep in mind that this is a gauge as defined in Chapter 7; the sums are conventional and their translation, even more so. But in the absence of other information relevant to assessment, this scoring can provide a convenient and useful gauge.

11.4. AN ILLUSTRATIVE EXAMPLE

We can illustrate the uses of this pedigree by a simple example: the introduction of hand-pumps for drinking water in rural areas in a less-developed Third-World country (Pacey, 1977). In this example, elements of several histories are combined, to provide a plausible reconstruction. Statistics on hand-pumps have been important for policy purposes in connection with the "Water Decade" mentioned above. For the availability of hand-pumps for drinking water would seem to be a reliable indicator for "access to safe drinking water". The vagaries and pitfalls encountered in the operationalization of this simple indicator can serve as an example of why an evaluative analysis, such as provided by the NUSAP system, is necessary.

Our story starts long ago, when a Ministry in a newly independent state was requested by some U.N. agency to report the number of pumps in rural areas. Lacking any means to study the phenomenon on the ground, the civil servants eventually located some numbers in files left behind by a previous regime. These were duly conveyed as 6,970. Were we advising a user concerning the reliability and history of this number, we would first need to elicit the information that enabled us to identify the appropriate modes. Supposing this done, we could express the quantity as:

6,970 : pumps : – : Poor : (Inert,Indir,No–c,None)
or   6,970 : pumps : – : 2 : (1,1,0,0).

We notice that even in such a case an advisor can, by using NUSAP, transmit an official statistic while expressing due warning on its quality.

Later a regional association organized a conference on "rural development"; all participants were to supply figures on hand-pumps. The new government of the country was not to be worsted by its neighbours; and so a report, complete with precise number, was drafted. Here the full NUSAP expression could be

15,432 : pumps : – : Poor : (Symb,Fiat,No–c,None)
or   15,432 : pumps : – : 1 : (1,0,0,0).

As the country developed in strength and institutional maturity, a national census of village pumps was organized. Enumerators visited every village, complete with instruction manuals and clip-boards. However, there had been an inadequate training and organization for the exercise; erroneous reporting and aggregation was common. We can express the result as

11,767 : pumps : – : low : (Conv,Dir,Evas,None)
or   11,676 : pumps : – : 6 : (2,3,1,0).

(We note that the Reliability Index here would be 6 or Poor; but still an improvement.)

A few years later, a similar census was organized, but this time it was designed to avoid the mistakes of the previous one. Some experts were brought in to advise on the organization; and the Army was used for the field work. The aggregated number reported in this case was $7,566 \pm 46$ or in the NUSAP system

7,566 : pumps : ± 46 : Good : (Sci,Dir,Obed,Reg)
or   7,566 : pumps : ± 46 : 10 : (3,3,2,2).

We notice that this is the first case in a which a spread entry is appropriate.

Knowing the sensitivity of an impending policy decision to the various properties of the number, we can adjust the NUSAP expression accordingly, to avoid hyper-precision. Turning to the quantifying end of the expression, we remark first that the Good strength in the assessment category entitles us to take the reported error term as the basis for the spread category. In this last case, the most rigorous expression for numeral, unit and assessment would be

75 : [100's pumps] : $\pm 1/2 \times 100$: Good : (3,3,2,2).

but for the sake of easy comprehension, this might be written for publication as

7,500 : pumps : ± 50 : Good : (3,3,2,2)

For the policy-maker, it could be sufficient to report

7,500 : pumps : ± 50 : Good

When the reliability is good, the precise pedigree is not always required; when it is poor or nonexistent, then a scrutiny of the pedigree may be the only way to make the information useful.

Eventually, the country joined those benefitting from international aid for development; hand-pumps were the subject of a special bilateral programme, largely administered through the private firms providing the equipment. At its formal conclusion, there was not much new to be seen on the ground. How to report that successful outcome which was so strongly desired? The technique was simple: all the items classed as "pump" on invoices and order forms were added to the results of the previous census. The supposed total provision of pumps was conveyed by a number whose full critical NUSAP expression would be

$$11{,}402 : \text{pumps}- : \text{Low} : (\text{Conv,Indir,No}-\text{c,None})$$
$$\text{or} \quad 11{,}402 : \text{pumps} : -: 4 : (2{,}2{,}0{,}0).$$

Let us briefly review what was understood as "pump" in each of the above cases, and consequently how the processing was done. In the very first case, "pump" was no more than an entry in an old file; hence we considered the mode of Definition and Standards to be inertia, while the number itself was at best an indirect estimate. In the second case, however, the symbolism swamped all other considerations; that to which the number referred was not pumps but a persuasive fiction. The first census enumeration was conducted on a common-sense basis; a "pump" was what each enumerator thought it to be, within a broad class of devices with a wide variety of degrees of operability. Hence we call that convenience. The second census, better organized, had some science in the definition of pump; those which were either not installed, or if installed, abandoned, were not to be counted in the enumeration. Thus at last some element of disciplined realism was brought into the process. But the temptations of progress soon worked their way in; and in the concluding report on the development contract, "pump" once again became a fictional unit of accounting. Since the entries on the invoices in this last case had some relation, however tenuous, to objects on the ground, we describe the mode as indirect estimate rather than fiat or guess.

To continue our narrative, we imagine that this last set of statistics for pumps (those derived from invoices and order forms) was sent to the regional headquarters of an international funding agency. There they were quickly recognized as nonsense. After a brief consultation with Ministry officials, some plausible substitutions were made. For this operation, the pedigree could be (Conv, Guess, Dial, None); thus dialogue can occur even when all other phases are vacuous or nearly so. But soon, this little helpful exercise was discovered by Head Office because of a "leak". Aware of their vulnerable position as international civil servants, they insisted that the numbers on the original submitted form be accepted, and so they were; the pedigree would then be (Symb, Indir, Obed, Ext). This presents us with an apparent paradox, for the "Reliability Index" in this case would be 9 or Medium. However, this case serves to remind us that the reduction of qualitative judgements to quantitative statements cannot be an automatic procedure. The present case is (hopefully) exceptional; and so competent experts noticing the unusual array of modes would already be alert to the possibility that simple addition of codes would yield a misleading figure in assessment. They would write Low for that category, so that users would be warned that something was amiss.

Returning to our story, a proper bilateral aid agreement was recently negotiated, for improvement of water and sanitation facilities in rural areas. Survey work was of the best international standard in its technical aspects. Field enumerators were well trained in the reporting conventions; but these were not well adapted to local social and cultural conditions within the country. Given their competence and commitment, they were able to make an informal accommodation between the international conceptual structures and local realities. Therefore, we can give a pedigree (Sci, Dir, Accom, Reg), in this case, or (3,3,3,2), with a Good assessment. When their report was received at the planning office of the bilateral aid agency, the experts were in something of a quandary. The programme was under close scrutiny, and the directors were concerned to avoid the suspicion of having understated the existing base-line provision so that their programme could later claim success. Hence they wanted a number that would be safe from understatement. Supposing that the original reported quantity was

$$11,300 : \text{pumps} : 5\% : \text{Good} : (a,b,c,d),$$

they could translate this to

$$\geq 10\tfrac{1}{2} : \text{K pumps} : - : \text{High} : (a,b,c,d)$$

The higher assessment rating could be based on the open-ended interval and on the coarse topology as expressed in numeral and unit respectively.

At the end of the programme, the agency was required to provide an independent evaluation of its results, before it could obtain a renewal from its sponsors. A highly expert independent team was sent to review the situation in the field. Their report was disturbing: many of the pumps seemed to be in a state of disrepair or even disuse. Moreover, in discussions with locals they found that many of the designs were inappropriate for local conditions; in one case the pumps caused marital strife in villages when wives refused to operate them on the grounds of their being so difficult as to cause ill health. Nor did general health improve, as in those conditions the pump's water was frequently as polluted as that from open sources. For their evaluative report, the definition of "pump" became crucial. Should they refer to all installed pumps, or to all those deemed operative, or only to those acceptable to the users, or (still more restrictive) those also providing safe water? By the time such problems of definitions were comprehended, the survey was underway, too late to be re-structured. In the report, the technical definition of pump was still used, but the tables were accompanied by extensive prose discussions of the limitations and pitfalls of the quantities reported. Here our pedigree reads: (Sci, Task, Dial, Indp), with High assessment.

In spite of these troubling findings, the programme as a whole was considered worthy of further support and improvement. The findings of the "task-force" survey were to be built into the specifications for the next phase of the programme. This proved to be not at all simple in practice. A number of conventions had to be agreed as to when a technically imperfect pump was to be counted as "operative". More difficult were such working definitions as those of "convenient" and "safe". As we have seen, some such terms depend on sociological and anthropological data; others on debateable scientific information. To achieve a realistic set of definable categories, the organisers found it useful to involve the local people themselves in some of the discussions. Hence the ongoing monitoring programme for the new project required a lengthy and expensive design process of its own. The pedigree for the information derived from that, is (Neg, Task, Dial, Reg), with High assessment.

In this hypothetical story of the evolution of hand-pump information, we have used pedigree as an analytical instrument in our discourse. In many of the cases in the story, users of the information would find then pedigree code sufficient for their diagnosis of the information. In some of the cases, the codes assigned could well be disputed; a resolution would then be achieved by a process of elicitation from the (fictional) producers of the information.

Traditionally, analytical philosophies have shown how elusive is the definition of everyday words like chair or table. The pedagogical purpose of the exercise is to enhance awareness about the problematic character of our "common-sense" knowledge. Here we have elucidated the variety of meanings of words that are only slightly more complex, as "seating" or "pump". Our purpose is to show how these terms are each constructed within a particular policy context, for the shape the relevant indicator in accordance with some agenda. The issues of uncertainty and quality are here not a matter for general conceptual analysis, but for a disciplined enquiry guided by a coherent conceptual system, as NUSAP.

11.5. INDICATORS: THE ELUCIDATION OF QUALITY

In general discussions of indicators, the term "quality" occurs frequently and in many contexts. For example, we hear of indicators of quality and of qualitative indicators, and even of quality of indicators. The similarities of expression, and the overlap of meanings, lead to some confusions, and conceals differences that are important for the proper definition and use of indicators.

Briefly, indicators of quality relate to the goodness (or otherwise) of some state of affairs relevant to policy. The quality might be of life, the environment, education, research or whatever. It might seem paradoxical to refer then to quality of indicators: but this is a judgement of the goodness of its performance of its function, that function being the representation of the goodness of something else. To distinguish between these two levels is essential for competent deployment and criticism of indicators. Finally, qualitative indicators invoke quality in the sense of non-quantitative. Of course, the boundary between quantitative and qualitative is vague, with gauges and taxonomies lying midway.

Before looking more closely to the problem of quality, let us clarify the distinction between *indicators* and statistical *indices*. A statistical index is, in its broadest sense, a measure of the magnitude of a variable at one point relative to its value at a base point. It is a statistic that may be gathered as a matter of routine, though it inevitably reflects the dominant conceptions of reality and of its representations. An indicator is used to gauge significant trends in some state of affairs. It may be a single selected index, or it may be compounded from several indices; it does not exist in isolation from its policy functions. The distinction between indices and indicators is illustrated in the etymology: the index is a pointer (as the index-figer or forefinger), whereas the indicator is the thing that points to some other thing. Many important indicators are called "indices", because they are routinely collected statistics; the distinction is one of function: thus the "Retail Price Index" may be used as an indicator for inflation.

We can provide examples for each of the different meanings of quality in relation to indicators. First, consider measures of "the quality of life". Because personal safety and security are so important to people, and also because a society should be seen to be well-organised, the "crime rate" is foremost among the indicators used to assess this aspect of quality. An increasing crime-rate is an indicator of a social pathology; but its meaning becomes a matter of competing realities. One may be of a decline in the moral standards of private life; the other, a decline in the fairness of social life. Do we need more of police, of parental discipline, of welfare or of jobs? Needless to say, any chosen indicator derives from indices that depend on categories and procedures which may be extremely artefactual and varied. It follows that any chosen indicator must reflect a particular conception of reality, and it will then be expected to confirm and reinforce it in the mind of all the public (including politicians and experts as well).

The "quality of an indicator", a judgement made on the goodness of its performance of its function, can be evaluated on technical as well as broadly political grounds. Thus "monetarist" economics, while a subject a strong debate in general, also encountered the difficulty of defining its crucial indicator of "money supply". In the U.K. both M_3 and M_0 were tried, but proved erratic and unreliable (*The Guardian*, 1985). The index M_3 survives as a statistic, but is no longer regarded as an indicator.

The problem of using indicators for the measurement of quality is well illustrated in science policy. Traditionally, scientific achievement was assessed mainly on numbers of publications. But this became too easily abused, in an age of proliferation of journals. Then ever more refined criteria and procedures were adopted, starting with the Science Citation Index. Since this covered only a small proportion of existing journals, it was accused of having a built-in bias; its criteria of "excellence" (based on citations) promoted English-language research at the expense of others, and also systematically excluded Third-World Science (Moravscik, 1985).

Another problem in the construction of indicators for the measurement of quality, is that of contradictory indices. We have another example in the science-policy field, in the definition of an indicator for the quality of research institutes in the Federal Republic of Germany (Sietmann, 1987a, 1987b). "Quality in Science" is no longer a debate among philosophers, but a struggle for survival of the fittest when resources are scarce. In the German case, several indices, each quite plausible, of scientific quality (as, Ph. D's, staff acting as referees, job offers, and overseas visitors) were compiled; but the various rankings for quality among the institutions were all different! Any chosen indicator would then represent a policy choice external to the assessments made by the indices.

Finally, we come back to quality of indicators, in the sense of how well they perform their functions. In indicators as in all quantitative information, quality is achieved not the elimination of uncertainty, but by its effective management. Forecasts are not assessed by their certainty, but by the justified confidence in their use. This is assured by a variety of means, including the ongoing processes of quality control at all levels, down from definitions of indicators at the policy level, through the technical work of construction of indices, and the operational level of data-collection and analysis, through to the work of review. In these various ways, quality is assured and uncertainty is controlled. The quality of indicators is enhanced when the characteristics uncertainties can be managed and communicated, as by NUSAP. Then policy decisions utilizing such enriched indicators will be based in a more sophisticated analysis of options and the balances among them; in these, uncertainty, and even ignorance, are effectively brought into the equation.

CHAPTER 12

MAPPING UNCERTAINTIES OF RADIOLOGICAL HAZARDS

An important example of policy-related research is radiological protection, as highlighted by the aftermath of the Chernobyl accident. Since radioactive agents had materials are now an inescapable part of our natural and technical environment, it does not need a major accident for radiological protection to be an urgent societal problem. The setting of standards for "permitted" or "tolerated" exposures is a continuous task, and one that is inevitably controversial. The processes whereby radionuclides, once released into the environment, enter the various ecological cycles and eventually cause human disease, are complex and only imperfectly understood.

As we have seen (Chapter 10), the study of radiological effects relies mainly on computational models and calculated data. With such a border with ignorance, this field cannot hope to reduce uncertainty to the level achieved in, say, experimental physics. The task, therefore, is to manage uncertainty to best advantage, so as to enhance and assure the quality of the information in these functions. Here we shall consider computational models for radiological effects that involve a small set of parameters and linear pathways. With the NUSAP approach, we can keep the different parts of the model distinct, and also map the relevant uncertainties so that their propagation through the models can be traced. By such means we provide an overall evaluation of the calculated model outputs; we also provide a simple and robust "mapping" technique for identifying critical uncertainties, so that further research for improving the model may be directed most effectively. Our technique is analogous to the "back of envelope" calculations that experts use for a preliminary estimation of quantities. It does not replace existing computational techniques for assessing uncertainties; but it complements them by providing a very quick and cheap method of estimation. It may be also be used for determining when the more complex techniques are genuinely worth applying.

The NUSAP scheme distinguishes between the traditional spread and an assessment of reliability which we may here call *strength*. With these two dimensions, we can map the uncertainties of the separate model parameters on an "assessment diagram". With appropriate conventions, this procedure enables us to trace the propagation of uncertainty through the model, and thereby to establish the overall reliability of the calculated model output. Since this procedure depends on an evaluation of the quantified model parameters, we must define a pedigree matrix for them. The quantification of the parameters is done on the basis of the input data; for them a different pedigree matrix is appropriate. In the following sections we will describe first the data-quality pedigree, then that for parameters, and finally the mapping procedure.

12.1. QUALITY OF RADIOLOGICAL DATA*

The data used for radiological models are enormously varied in provenance and quality. There are many individual radionuclides to consider, and very many contexts of their activity. For example, the "uptake" by crop plants will depend strongly on the radionuclide, the type of radioactivity, as well as on soil type, moisture, rainfall, and the type and variety of the plant. This great variety has several effects on the quality of the data. First, for any given context, there are liable to be only a few studies at most, and their quality assurance may be unconfirmed. Also, the relevance of a particular data item to the problem at hand is a matter of degree, to be evaluated by a judgement of the similarity of the two contexts. This involves what we have called analogical reasoning, clearly a weaker form than deductive or inductive. Such structural weakness as these are not usually reflected in the precision of the numbers that are recorded as the entries in a radiological data banks.

The NUSAP expression for radiological data consists of the standard five categories, preceeded by a set of *keywords*. These provide descriptive terms for identifying the circumstances of production of the item. For example, the set of keywords for an entry giving the "soil retention factor for Oxfordshire clay" could simply be the groups of the first three and the last two words. Different entries will have different number and type of keyword descriptive terms. They will typically include information about the "population" to which the data relates, where population is used in a generic sense to include people, cows, grass, blocks of soil, raindrops, radionuclides, or any other subjects of relevance to radiological modelling. The descriptive terms may also include information about the geographical and temporal specificity of the data, and any other relevant identifying characteristics. Some keywords, or descriptive terms, will give precise and unambiguous reference to a particular subject of interest: those designating a particular radionuclide or a particular field site are of this kind. Others will be less determinate: terms such as people, grass or soil are examples, their meanings having a degree of openness, often referring to an unidentified average or "ideal type" of their class.

The pedigree matrix for radiological data is composed of three phases: *Type*, *Source* and *Set-up*. As with other pedigree matrices, there is a set of normatively ranked evaluative modes for each phase, in whose terms judgements of quality can be derived. The pedigree matrix is displayed in Table IX.

The phases, with their corresponding modes, are defined as follows.

- **Type**. This phase represents the relative strength of the interferences by which the data are derived. The modes for this phase are:
  - *Constants* (physical or mathematical).

---

*The following sections are based on research done in collaboration with S.M. Macgill. See Funtowicz, Macgill and Ravetz, 1989a,b,c.

TABLE IX
The pedigree matrix for radiological data-entries

Code	Type	Source	Set-up
4	Constants	Reviewed	Universal
3	Deduced	Refereed	Natural
2	Estimated	Internal	Simulated
1	Synthesized	Conference	Laboratory
0	Hypothetical	Isolated	Other

- *Deduced*. Data sources are empirically rich. These may be either direct measurements; or validated numerical operations on measurements, or they may be derived from established theories.
- *Estimated*. These data are inferred from sources of general relevance. Judgements need to be made as to reliability of particular values.
- *Synthesized*. Only weak data sources are available. Models must be used for production of input data, and these cannot be validated or tested.
- *Hypothetical*. Data are "assumed". They are based on pragmatic considerations, conjecture and perhaps even on speculation. The processes they depict have not been measured and perhaps cannot even be observed.
- **Source**. This phase represents an evaluation of the reliability of the origin of the data in terms of the quality assurance of particular sources. It makes explicit the traditional scholarly evaluations of peer-review. The modes for this phase are:
  - *Reviewed*. The data entry is the outcome of a review procedure across a wide range of sources. It may be obtained from a review article in the literature, or elicited from experts with the requisite knowledge.
  - *Refereed*. The data entry is obtained from an article in a reputable scientific journal, for which a full refereeing procedure can be assumed.
  - *Internal*. The data entry is an in-house result which has been given institutional scrutiny.
  - *Conference*. The data entry was presented in conference proceedings. It has passed the quality control of the conference organizers, but has not necessarily been subjected to specific vetting, nor to a full refereeing procedure.
  - *Isolated*. Information which has not undergone any known quality control procedure. Neither the publication, the institution, nor the author has standing.
- **Set-up**. This phase describes the conditions of the study which yielded the data entry. The modes for this phase are:
  - *Universal*. The study conditions are valid for every single situation of interest (e.g., radiological decay constants).
  - *Natural*. The study conditions have no known significant differences from

natural situations (e.g., measurements of radiation intake by cattle feeding under normal or representative conditions).
- *Simulated.* "Artificial" aspects are imposed for the achievement of greater control (e.g., feedlot experiments with cattle), or for generality (e.g., averaging over several different study conditions).
- *Laboratory.* Experiments and simulations which are significantly different from natural conditions.
- *Other.* Miscellaneous studies.

We notice that laboratory is here a mode with a very low rank; this is in contrast to its usual evaluation. The radiological models are intended to describe processes in the natural environment, which in general will be very different from those in the artificial conditions of the laboratory. For some purposes, data which is derived from several rather uncertain sources in nature may be more useful to the modeller than a precise result obtained in the laboratory. In such cases, there is obviously a need for highly skilled judgements of quality; this hierarchy of modes is designed to guide and support them.

We provide two examples of the NUSAP representation of data entries. These will be used later as inputs for a food-chain model.

λ_r: radioactive decay constant for caesium 137,
   Keywords : Cs 137, radioactive decay constant
   Numeral : 0.023
   Unit : l/year
   Spread : –
   Assessment: high
   Pedigree : (4,4,4)

I: milk intake rate for individuals
   Keywords : Average milk consumption, total UK population
   Numeral : 150
   Unit : kg/year
   Spread : ± 50
   Assessment: medium
   Pedigree : (2,2,2)

The difference in quality between the two data entries is immediately apparent from their pedigree coding. The user is given a clear cautionary signal about the quality of the milk intake data. The data entry was estimated from institutional data on milk consumption, and not obtained experimentally. It does not carry the authority of a fully refereed result, but is only an in-house, internal product (National Radiological Protection Board, 1987).

The keywording and NUSAP representation of data entries can be updated in line with new information and developments. In the case of the pedigree ratings, for example, a code that reflects a data source as an internal publication will be altered following acceptance in a recognized journal; and it will be due for revision again on any subsequent achievement of the status of "authoritative review". The codes in all cases should generally reflect where the state-of-the-art has reached, but should remain stable for a reasonable time in order to preserve continuity. The systematic coding of data by the NUSAP system will provide users with clear signals about the quality of data available for radiological modelling. Such codings could be done routinely, enabling judgements about data quality to be preserved (and updated, as appropriate) on a coherent basis. This can be the foundation for the subsequent construction of a full "information system" of critically evaluated data entries. The NUSAP system can also be used for developing the skills for making evaluative judgements on the materials used for radiological models.

12.2. QUALITY EVALUATION OF RADIOLOGICAL MODEL PARAMETERS

In this section we are concerned with models specified by an algebraic expression consisting of parameters connected by operations and functions. The parameters correspond to the objects and processes in the natural system being modelled; in the algebraic expression they appear as variables, taking on particular quantified values when there is to be a calculation. The quantification of parameters is not a simple operation, as there must be a selection and interpretation from among the available data entries that are relevant to it. The problem of the quality of those data entries in this context of use introduces a special sort of uncertainty into the calculated model output. This is not described by the existing techniques of sensitivity and uncertainty analysis; for these are restricted to calculating with the spread of the parameters. By extending the NUSAP representations to quantified parameters, we can assess and describe their quality in relation to that of the data from which they are derived. On that basis we can incorporate their uncertainties into the mapping procedure, and thereby obtain a quality assessment for the calculated model outputs in every case.

Evaluations of the radiological model parameters can be based partly on the same aspects as evaluations of the radiological data, since the quantitative estimates of model parameters will themselves be derived from the available data. Criteria for assessing the quality of data can therefore be expected to be important in evaluating the quality of model parameters. But other aspects must be added, for a single radiological model expressed algebraically may have a variety of interpretations, according to its intended use. These uses include: routine monitoring of small emissions; estimations of dosages from a large releases in emergency situations; setting of threshold values in regulatory standards; application to locations or populations that are either general or specialized; or even studies in the art of modelling. Depending on a model's use, its parameters may be interpreted differently, quantified differently, and even derived from different data sources. The evaluation of quality of model parameters must be correspondingly flexible.

As an illustration, we analyze the 'pasture-cow-milk' model (National Radiological Protection Board, 1986); which consists of the following equations:

$$\text{Dose (Sv)} = H_e \times I \times C_m \tag{1}$$

where

H_e is the committed effective dose equivalent per unit intake, Sv/Bq;

I is the milk intake per individual, kg/y; and

C_m is the time-integrated concentration of radionuclide in milk, Bq × y/kg.

This represents the dose received by a person drinking the milk; we notice that it is in Sieverts, or Sv, the unit of radioactivity acting on the body, while Bq is the unit of radioactivity in the source. The dimensions of the three terms combine to give Sv, as (Sv/Bq) × (kg/y) × (Bq × y/kg).

For the radionuclide in milk, we have the equation

$$C_m = F_m \times O_f \times C_a \tag{2}$$

F_m is the milk transfer factor, d/kg.

O_f is the cow's intake rate of grass, kg/d; and

C_a is the time integrated concentration of radionuclide in pasture grass, Bq × y/kg.

Finally, the radionuclide in pasture grass is given by the equation

$$C_a = D_i \times \left[\frac{R}{\lambda_w \times Y} + \left(\frac{B_v \times w}{d \times \rho} \times \frac{1 - \exp(-\lambda_s \times t)}{\lambda_s} \right) \right] \tag{3}$$

This equation expresses a model with two parallel pathways, through grass and roots, whereby radiation becomes ingested. The first term represents the radioactivity on the grass. It is assumed that because of the rapid weathering (half-life is days), this can be given as a total-dose, time-independent form. Hence it appears as an integral to infinity of a decaying radioactive source, as is shown by the parameter λ_w in the denominator. The second term represents the radiation entering through the roots, by uptake from the soil. Here the decay rate is a compound of that of the radioactive substance itself and of the displacement of soil, as expressed in the "soil retention" parameter with dimensions y/m. Because this rate-constant is much smaller than the other, a time-dependent term is included. This will provide interesting variations in the calculated model output.

Here the parameters are as follows (4):

D_i is the initial, Bq/m²;

R is the interception factor for pasture grass;

λ_w is $(365 \times \ln 2)/T$, where T is the weathering half-life in days;

Y is the grass yield = 0.1 kg/m²;

B_v is the root uptake factor, (Bq/kg wet weight grass)/(Bq/kg dry wet soil);

w is the ratio of total weight of plant to wet weight;
d is the depth of rooting zone, m;
ρ is the density of the soil, kg/m^3; and
λ_s is the rate constant for loss from soil, 1/y.
  $= (d \times r_s)^{-1} + \lambda_r$, where r_s is the soil retention, y/m, and λ_r is
the radioactive decay constant, 1/y.

The parameters as quantified are (5):

	Best estimate	Range (or spread)
H_e	1.8E–8	(4.16E–9, 2.08E–8)
I	150	(100,200)
F_m	7E–3	(2.5E–3, 2.5E–2)
Q_f	12	(10, 18)
R	0.25	(0.1, 0.4)
T	15	(5, 30)
B_v	2E–2	(5E–3, 8E–2)
d	0.15	(0.1, 0.2)
ρ	1.6E3	(1.3E3, 1.9E3)
r_s	4E2	(5E1, 1E3)

The pedigree matrix for model parameters is based on the three phases of the previous pedigree matrix, Type, Source and Set-up (Table IX), together with two new phases, *Relevance and Processing* (Table X). The three phases of the previous pedigree matrix are repeated here, so that quality evaluations about data can be carried over to the corresponding model parameters. The fourth and fifth phases relate the quantified parameters back to the data entries on which they are based.
Relevance is self-explanatory; and Processing describes the adjustements made to data entries as they are used in the quantification of parameters in particular contexts.

Relevance and Processing, with their respective modes, are defined as follows.

– **Relevance.** For any given parameter this phase reflects the degree of correspondence between the parameter as interpreted in the particular context

TABLE X

The relevance and processing phases of the pedigree matrix for radiological model parameters

Code	Relevance	Processing
4	Full	Confirmed
3	High	Aggregated
2	Good	Extended
1	Medium	Accepted
0	Poor	Copies

of the model as applied, and the data entries from which its quantified value is derived. There may, or may not, be data entries of good relevance available for the particular quantified parameter. The modes for this phases are given as a normative scale (from High to Poor), rather than in terms of a semi-formal categorization of the differences between parameters and data entries. We have adopted this procedure because the differences are so various that no simple and robust notational scheme would encompass them. Using NUSAP to guide expert judgement rather than to replace it, we prefer to keep the modes simple, and to incorporate the complexity of the judgement into an elicitation procedure. For this we compare the "data as given" and the "parameter as needed", in terms of the Set-up and the Keywords, in a simple display. By comparison of the paired entries, the user can form a judgement of the degree of relevance of the two, and define the mode accordingly. We shall illustrate this by examples below.

- **Processing**. This phase conveys the nature of the operations performed on data entries in order to produce quantified parameters. Data entries are not necessarily adopted as given; their spreads may be expanded to make prudent allowances against uncertainty (trade-off), or parameters may be compounded from several data entries. The modes of this phase are:
  - *Confirmed*. There is an explicit verification procedure about the suitability of the data entries.
  - *Aggregated*. The quantification procedure involves the aggregation of two or more data entries.
  - *Extended*. The quantification procedure involves a spread-assessment trade-off.
  - *Accepted*. Data entries are judged to be acceptable for the quantification procedure.
  - *Copied*. Quantitative information for parameters are merely copied, without much reflection, from the data entries.

12.3. ILLUSTRATION OF PEDIGREE RATINGS FOR MODEL PARAMETERS

We will illustrate the pedigree for model parameters, using three parameters of the pasture-cow-milk model: λ^r (radioactive decay constant), I (milk intake rate) and r_s (soil retention factor). We consider three different applications of the model:

- Problem 1: Uncertainty analysis of calculated model outputs.
- Problem 2: Dose estimation: routine case, general UK population.
- Problem 3: Dose estimation: children in a specific location.

These different problems are chosen to show how judgements on quality of parameters should take account of the context of application of the model. They are thus inherently problem-dependent, provinding a good example of this aspect of policy-related research.

The radioactive decay constant λ_r is very simple. It would be given high codes for all pedigree phases, reflecting its universal strength. The pedigree of λ_r is (4,4,4,4) for all three cases.

For our example we consider the second parameter, the milk intake rate I. Its numerical value is given in (5). As we discussed before, codes of 2 for Type and 2 for Source would be appropriate, given that the data were estimated from institutional data on milk consumption. For Relevance, we proceed by exhibiting the Set-up and keywords entries for the model parameter, as appropriates for the first problem (uncertainty analysis). We record corresponding information for the data entries from which the parameter is to be quantified. This gives:

I	Set-up	Keywords
Data as given	Simulated	Average milk consumption rate, total UK population
Parameter needed	Simulated	Average milk consumption rate, total UK population

The details are identical, therefore the Relevance code is 4.

We can repeat the procedure for Problems 2 and 3. For the second (routine dose estimation), there is a change in Set-up, where the mode for the parameter as needed is Natural. We also record in keywords that the parameter will be used for a general" UK population. We notice here that the milk consumption data is to be applicable to all UK individuals (not just "average" ideal individuals). There is a clear decrease in Relevance from the first problem. We might judge the correspondence nevertheless as High, and give it a code 3.

For the third problem (dose estimation, children, specific location) we have:

I	Set-up	Keywords
Data as given	Simulated	Average milk consumption rate, total UK population
Parameter needed	Natural	Distribution of milk consumption rate, local children

There is a marked difference between the two sets of details, which should be reflected in a lower code for Relevance, say 2.

Still in the second example, the parameter I, we can now consider the phase Processing of the pedigree matrix. Taking the range as given in (5), we would give a code of 2, reflecting that a trade-off was made. A deeper discussion here would consider each problem separately, as we have done with the previous phase.

For our third illustrative example, we consider the parameter r_s (soil retention factor). The information for Relevance, first problem (uncertainty analysis), may be displayed as follows:

r_s	Set-up	Keywords
Data as given	Natural	Soil retention factor, USA location specific
Parameter needed	Simulated	Soil retention factor, UK applicable

The correspondence between the keywords for the data and the parameter are reflected in a relevance code of 3: it is reasonable to use data from a Natural Set-up for a Stimulated uncertainty analysis. However, since it is not derived from UK data, it is appropriate not to code it as 4.

For the second problem (routine dose estimation) the only change is in the keywords, where instead of "UK applicable" we have "UK general". It is judged that the Relevance code is still 3. Finally, for the third problem (dose estimation, children, specific location), although the Set-ups for data and parameter, are now the same (Natural), the keywords will record the fact that what we need is a parameter appropriate for "UK location specific". The applicability of USA location specific soil data to a model for a specific UK location will depend crucially on the two locations in question. It is not difficult to imagine some cases where the correspondence would be very good, and this would be reflected in a Relevance code of 4. But it is not difficult to imagine a case where the correspondence would be very poor indeed. To allow for all these possibilities we use the flexible notation j ($j = 1,2,3,4$), to represent the Relevance code in this case.

Finally, we consider the Processing phase for this quantified parameter. As with I, we do not consider each problem separately, but take the single range of estimates as given in (5), and assign a code of 2, reflecting that a trade-off was made. A summary of the derived pedigree codings for each of the illustrations discussed above is given below:

Parameter	Problem 1	Problem 2	Problem 3
λ_r	(4,4,4,4)	(4,4,4,4)	(4,4,4,4)
I	(2,2,4,2)	(2,2,3,2)	(2,2,2,2)
r_s	(2,3,3,2)	(2,3,3,2)	(2,3,j,2)
			where $j = 0,..., 4$

We notice how in the second and third examples, the rating for Relevance decreases as the problems become more specific.

12.4. PARAMETER UNCERTAINTY AND MODEL RELIABILITY

All the previous studies on the propagation of uncertainties through radiological models have used only spread (see, for example, Eisenbud, 1987), neglect-
ing the reliability, or strength, which NUSAP expresses in its assessment
category. With these two dimensions, we are able to produce a mapping which
displays the propagation of uncertainties from the model parameters to the
calculated model output (Funtowicz et al., 1988).

Calculated model outputs will not necessarily be vulnerable to a large
spread in quantified parameters. This is because the mathematical structure
of the model may be such as to swamp the spreads of such parameters, and
even their quantified values, as in minor parallel pathways or in a compart-
mental model. In this case, a quantified parameter will contribute relatively
little to the total spread of the calculated model output, no matter how large
a spread it has. When we consider the relative contribution of parameter
spreads to the total spread of the calculated model output, we will need to
make a special calculation in each case.

Turning now to the more qualitative aspect of the uncertainty of model
parameter inputs, we also find that calculated model outputs will not necessarily be vulnerable to low strength in parameters. Much will depend on the
mathematical structure of the model, and any potential adverse effect may
well be swamped. The danger cases for the model are "low strength" parameters which contribute a relatively high proportion to the total spread of the
model. Point A in Figure 10 illustrates such a case. For comparison λ represents a strong parameter whose relative contribution to the total spread is low.

The A and λ in Figure 10 could be parameters in a model of a radiological
decay process, where uptake is of the form

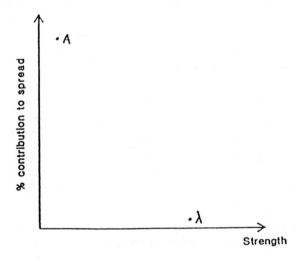

Fig. 10. Basic assessment diagram.

$$A \times [1 - \exp(-\lambda t)]$$

We can suppose that λ is well known, and that t can be determined at will. But A, representing various transfer functions, may be not at all well known. The model is very vulnerable to A, but not at all vulnerable to λ or to t. Any evaluation of the reliability of the calculated model outputs should accordingly focus on the parameter A. Its position in the diagram, reflecting spread and strength, leads to an evaluation of the model as "poor".

The interpretation of more general cases of spread and strength is accomplished by the Assessment Diagram of Figure 11. This has been designed to suggest that the robustness of calculated model outputs to parameter uncertainty can be good even if parameter strength is low, provided that relative contribution of that parameter to that model spread is also low. In this situation, our ignorance of the true value of a parameter is of little consequence: the parameter spread has a negligible effect on calculated model outputs. Alternatively, calculated model outputs can be robust against parameter spread even if the relative contribution of that parameter to total model spread is high, provided that parameter strength is also high. As in Figure 11, the danger cases for the model are those where parameters of relatively low strength contribute a relatively high proportion of the total model spread.

These considerations lead us to suggest that parameters which lie anywhere

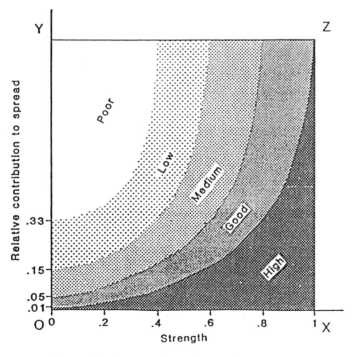

Fig. 11. Assessment diagram, showing zones for evaluation of calculated model output.

along lines OX and XZ are "safe". As we move away from those lines of safety, model reliability can be jeopardized by the increasing amounts which parameters contribute to model spread (upwards), and by decreases in parameter strength (to the left). The further away we are from point X (exceptionally high parameter strength and negligible parameter contribution to model spread), the greater the loss of safety. Throughout the diagram, increases in strength can compensate for increases in parameter contributions to model spread, but only to some extent. Figure 11 reflects these considerations in the differently shaded regions of decreasing model reliability, as we move from the lower right to the upper left. We have chosen intersections with the vertical axis at 1 %, 5 %, 15 % and 33 % for the regions of high, good, medium, low and poor reliability respectively. These percentages are often invoked in statistical confidence judgements. We have chosen intersections with the line YZ (where a single parameter accounts for all of the spread in a model), corresponding to parameter strengths gauged as ≤ 0.4, ≤ 0.6, ≤ 0.8 and ≤ 1 for the regions of poor, low, medium and good reliability, respectively.

The diagram as given is a convenient heuristic tool. It has been drawn in a way which balances the relative sizes of the different regions, bearing in mind also that for any given model, there can be at most one parameter in the upper half of the diagram (there cannot be more than one parameter that contributes more than 50 % of the spread in a calculated model output). The reliability of calculated model outputs in relation to parameter uncertainties can be evaluated from Figure 11 by taking the position of the worst case parameter, that is most dangerous to the model. This weak-line guideline, based on the principle that one weak parameter is enough to vitiate the entire calculated model output, will yield a prudent assessment of the reliability of calculated model outputs.

In order to make use of Figure 11 for evaluating the reliability of calculated model outputs, we need gauges for:

- Parameter strength (for the horizontal axis)
- Relative contribution of individual parameters to the total spread of the calculated model (for the vertical axis).

It is quite simple to produce a gauge for translating the quantitative judgements of reliability (or strength) of parameters onto the scale of (0,1). For this, we may use the scoring procedure on the pedigree codings of each of the quantified parameters. With five modes in each phase, we code them from 0 to 4, add the separate codes, and compare this sum to the maximum possible sum (here 16). Thus we obtain a normalized gauge for strength, always understood to be on a very coarse topology. (See Chapter 9.)

12.5. PARAMETER CONTRIBUTION TO MODEL SPREAD

In this section we discuss ways to derive a measure for representing the relative contribution of individual model parameters to the total spread in calculated model outputs. The first is an algebraic method, which we have developed as an extension of the approach used by Eisenbud for models having a relatively simple mathematical structure. The second, numerical, approach is suitable for more complex models, and would make use of results of uncertainty and sensitivity analyses, employing computational methods.

Suppose first that we have a model M consisting of two parameters A and B, so that:

$$M = A \times B$$

By simple calculus we can obtain the total differential

$$dM/M = dA/A + dB/B \tag{6}$$

We notice that this equation expresses how any small proportional change in M is simply the sum of the corresponding proportional changes in A and B. The integral of dM/M is simply the sum of the (separate) integrals of the two terms on the right hand side. Suppose now that A and B have given spreads (or ranges of variation); then integrating (6), we have

$$\log[M_2/M_1] = \log[A_2/A_1] + \log[B_2/B_1]$$

We may rewrite this simply as

$$\Lambda(M) = \Lambda(A) + \Lambda(B) \tag{7}$$

where Λ denotes the logarithmic spread, or range. This representation is particularly useful when the spread of a parameter is large in relation to its best-estimate value.

The relative contribution ρ of the spread of any parameter to the total spread in calculated model outputs, follows immediately as

$$\rho(A) = \Lambda(A)/\Lambda(M) \tag{8}$$

$$\rho(B) = \Lambda(B)/\Lambda(M) \tag{9}$$

We notice that the same calculations apply if A and B are each not single parameters, but rather groups, or blocks, of parameters. These may describe stages of linear pathways, or compartmentalized models.

Suppose now that a model, or part of it, has two parallel pathways. The amount of the substance in question arriving at their common endpoint equals the sum of their outputs. Then we have:

$$M = C + D$$

If the spreads of C and D are small, and are represented additively, then a straightforward calculus can be involved as before. But if they are large, and represented multiplicatively or logarithmically, we cannot directly calculate their relative contribution to the total spread in the model. Instead we proceed as follows:

$$dM = dC + dD$$
$$dM/M = dC/(C + D) + dD/C + D)$$
$$dM/M = (dC/C)/(1 + D/C) + (dD/D)/(1 + C/D)$$

The terms $1/(1 + D/C)$ and $1/(1 + C/D)$ are in effect the "elasticities" of M with respect to the parameters C and D respectively. Elasticity, a term used in economics, defines the proportional change in an aggregated entity, such as M corresponding to small proportional change in one of its component inputs, such as C or D. In this case, magnitudes of the elasticities depend on the ratio of the point-values of the two parameters.

In this case we cannot perform a straightforward integration, since the variables C and D are not separated out into their respective differentials. However, there is one case of interest when we can make approximations that enable an integration to be done. This is when the point-values of C and D are very different in magnitude. Suppose C is much larger than D. Then C/D is much larger than unity, and D/C is much less. The elasticity terms are then simplified, so that

$$1/(1 + D/C) \sim 1$$
$$1/(1 + C/D) \ll 1$$
$$1/(1 + C/D) \sim D/C$$

Now the differential form can be given the approximate equation

$$dM/M = dC/C + (D/C)dD/D$$

Given that the point-value of C is much greater than that of D, then for purposes of this rough calculation we may assume that the ratio of their values is effectively constant, throughout the range of variation of C and D. Then we have an exact differential form as in the first case, and by integration as before, we have

$$\Lambda(M) = \Lambda(C) + (D/C) \times \Lambda(D)$$

We can now calculate the relative contributions to total spread, as before, with the approximate equations

$$\rho(C) = \Lambda(C)/\Lambda(M) \tag{10}$$
$$\rho(D) = (D/C) \times \Lambda(D)/\Lambda(M) \tag{11}$$

These shows in a simple mathematical form, how the spread in a less important parallel pathway makes a correspondingly smaller contribution to the total, by the ratio D/C. In this case, a parameter may have a very large spread, and yet still not weaken the calculated model output to any significant degree.

Formulae (8), (9), (10) and (11) for determining the relative contribution of parameters to the total spread in calculated model outputs, can be used to calculate appropriate vertical ordinates on Figure 11. Visual inspection of the figure will then yield an evaluation of the reliability of the calculated model output, as required.

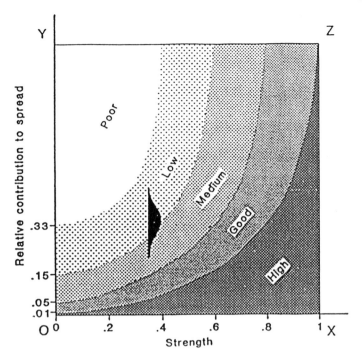

Fig. 12. Assessment diagram, showing distribution for calculated sensitivity analysis.

While the above algebraic analysis provides a quick and simple means of calculating the relative contribution of individual parameter, or parameter blocks, to total model spread, their use will be restricted to models with a relatively simple structure. For more complex cases, and where there is strong interdependence between parameters, the algebraic approach may no longer apply. Instead, it may be necessary to resort to sensitivity of uncertainty analyses using computers, for estimates of the contribution of parameters to the spread of calculated model outputs.

In numerical sensitivity and uncertainty analyses, different model runs are generated using different combinations of parameter input values. The output is a distribution on the total spread of the calculated model output in terms of different possible combinations of parameter input values. A single parameter is likely to be found to contribute a different percentage to the total spread in calculated model outputs in different model runs. For example, a numerical uncertainty analysis entailing 50 different model runs will yield, for each parameter, a distribution of 50 percentage contributions to the total spread in the calculated model output. For each parameter, these individual distributions can be superimposed onto the Assessment Diagram as representations of the relative contribution of individual parameters to the total spread in the calculated model, output (Figure 12).

Visual inspection of the completed Assessment Diagram will yield, as before, an overall evaluation of the reliability of calculated model outputs. Unlike with the earlier algebraic approach, here we do not have the precision of individual points in the diagram on which to base an overall assessment. Rather, there is a series of distributions, each of which may span two or more zones in the diagram. It would be a false precision to condense these distributions in order to construct an apparently more definite basis for assessment. All we can do is to make a judgement on the basis of the given display, including its scatter. We may well deem it reasonable to neglect thin "tails" of a distribution, but otherwise we should be guided by a "weak link" rule in arriving at an overall evaluation of model reliability.

12.6. ILLUSTRATIVE EXAMPLE

We provide an illustration of the use of the Assessment Diagram with the "pasture-cow-milk model". This includes two parallel pathways, grass and roots, through which radionuclides can be transferred into cows, and a downstream pathway to a milk-drinking population. The ten parameters in the model can accordingly be grouped into three blocks: G (for grass); R (for roots); and D (for downstream); with the model as a whole of the form

$$\text{Dose} = (G + R) \times D$$

We can consider the output from such a model for elapsed times of a week (w), a year (y) and a century (c). In view of the relatively simple algebraic structure of the model, the relative contribution of each block of parameters to the total spread in the model can be calculated with the algebraic approach described above. The model can be represented as

$$M = (G + R) \times D$$
$$\Lambda(M) = \Lambda(G)/(1 + R/G) + \Lambda(R)/(1 + G/R) + \Lambda(D)$$

To estimate the relative contribution of each block of parameters to the total model spread, we must obtain the point-value of R/G. The block R contains a time-dependent term, $[1 - \exp(-\lambda_s \times t)]$. For t very large, this is close to 1; for t very small (1 week = 0.02 year, or even 1 year), it is close to $\lambda_s \times t$. For our calculation we use the best estimate values of the parameters, given in (5). Taking values of t of 1 week, 1 year and 100 years, we have elasticities as follows:

Time	Grass elasticity	Roots elasticity
t	$1/(1 + R/G)$	$1/(1 + G/R)$
1 week	0.95	0.05
1 year	0.8	0.2
1 century	0.015	0.98

TABLE XI
Log-spreads for the parameter blocks

Parameters/blocks	Parameter log-spread	Block log-spread
Downstream		
H_e	5	
I	2	
F_m	10	19
Q_f	2	
Grass		
R	4	
T	6	10
Roots		
B_v	16	
d	2	
	1.5	25.5 (or 23)
λ_s	3	
$[1 - \exp(-\lambda_s \times t)]$	3 (or 0.4, for t = century)	

These fit with common-sense of the processes in the field. The radiation on the grass is strong, but weathers quickly; its contributions to the total dosage comes at the beginning. Hence w is large, and $\exp(-\lambda_w t)$ is negligible for times greater than a few weeks. The time-dependent term is therefore omitted from the model for greater elapsed times. The pathway through the roots is weaker, but the radiation decays slowly. Hence that pathway becomes more significant, and eventually dominates.

The logarithmic spreads of the individual parameters (see (5)), give the logarithmic spreads for the parameter blocks (Table XI). These block log-spreads are then combined with the elasticities to give the total log-spread of the calculated model outputs:

For t = 1 week $M = 19 + 10 \times 0.95 + 25.5 \times 0.05 = 29.8 = 30$
For t = 1 year $M = 19 + 10 \times 0.8 + 25.5 \times 0.2 = 32.1 = 32$
For t = 1 century $M = 19 + 10 \times 0.015 + 23 \times 0.98 = 42.2 = 32$

From here the relative contributions of each of the three blocks of parameters to the total log-spread follow immediately:

Time	Downstream	Grass	Roots
1 week	19/30 = 0.63	9.5/30 = 0.31	1.3/30 = 0.04
(rounded)	= 0.65	= 0.30	= 0.05
1 year	19/32 = 0.59	8/32 = 0.25	5.1/32 = 0.16
(rounded)	= 0.60		= 0.15
1 century	19/42 = 0.45	0.15/42 = 0.004	23/42 = 0.55
(rounded)		= 0.005	

The significance of these values becomes apparent when the parameter blocks are plotted on the Assessment Diagram. For this we need, in addition to the information above the ratings for the strengths of the three blocks of parameters. A fully accurate plot would require complete pedigree codings for each of the ten parameters individually; here we estimate strengths for the three blocks, as follows:

– Downstream: Although the spread is large, F_m and Q_r are well studied, and I is reasonably estimated. We let the average rating for the block be 3 (out of 4), giving 3/4 as the strength.
– Grass: Here the parameters seem to be estimated at best (Y, λ_w), or merely assumed (R). We let the rating be 2, giving 1/2.
– Roots: Most of the parameters are strong, and the weak parameter r_s is absorbed into λ_s; but B_v dominates the block, with its very large spread and presumed low strength. We let the average rating be 1 or 1/4.

Figure 13 shows that for times of one week and one year, all the points lie

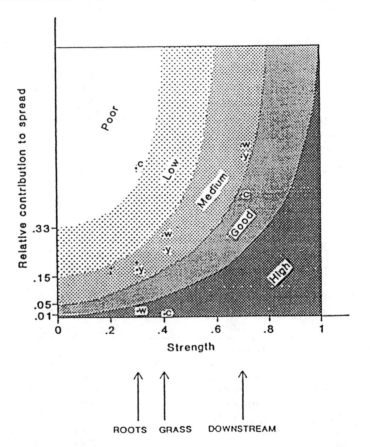

Fig. 13. Assessment diagram, applied to a radiological model.

within the zones of medium, good and high reliability. Thus, in spite of the considerable spreads in some of the parameters, the blocks are of reasonable strength, and the model as a whole is quite well balanced in the relative parameter contributions to spread. Only in the case of $t = 100$ years is there a significant imbalance; the weakness of the roots block is displayed by a point just inside the zone of poor reliability. By the prudent "weak link" principle, the calculated model output is then given a poor (or perhaps "low to poor" reliability rating.

By this illustration we see how the assessment mapping integrates the qualitative with the quantitative aspects of uncertainty in a single perspicuous display. By its means we can trace the propagation of uncertainty through the model, and thereby evaluate the uncertainties in calculations of radiological hazards, clearly and effectively.

CHAPTER 13

FURTHER APPLICATIONS OF THE NUSAP SYSTEM

In this chapter we provide further examples of the flexibility of the NUSAP system in policy-related research. As we have seen in the last two chapters, each area of research has its own problems of uncertainty management. For each one an appropriate technique can be developed within the general approach of NUSAP. A great variety of special forms can be designed and deployed.

The basic NUSAP scheme provides a coherent framework for all the particular notations (including pedigree matrices). It also provides guidelines for elicitation, to assist in the design of new notations. Since different areas of research will have different critical distinctions among their characteristic uncertainties, different pedigree matrices within the overall family will need to be designed. We have already seen this, in the cases of statistical information and radiological protection. In such work, we always keep in mind that NUSAP is above all a simple and robust tool. It fosters the skill of working at an appropriate level of precision, for both calculation and expression.

13.1. AN ARITHMETIC FOR ASSESSMENT*

We have already shown (Chapter 3) that when arithmetic is applied in contexts outside pure mathematics, the standard rules must be modified. In every case, an artefactual arithmetic must be developed. The more quantitative categories of NUSAP are very useful in clarifying these artefactual arithmetics that are already being used, some self-consciously (significant digits) and others not ("fossils joke"). With NUSAP we have been able to extend arithmetical methods to more qualitative judgements as well. In connection with statistical information (Chapter 11) we showed how, in the absence of other information on reliability we could use the pedigree codes as the basis for a very simple scoring system, providing a gauge for the assessment category called Reliability Index.

In connection with radiological protection (Chapter 12), we could map uncertainties by employing the interaction of spread and pedigree. The spread of each parameter appeared as a relative Contribution to Total Model Spread; and the contribution of pedigree was through Strength, derived from a scoring procedure on the pedigree codes. Each parameter (or block of parameters) was then described by a pair of attributes in coarsely quantified form. These

*The following two sections are based on research done with R. Costanza.

defined a point in an Assessment Diagram, located against a background of zones denoting reliability. Each parameter-point was thus identified for its significance for the overall reliability of the model. Thus, in spite of the deliberately coarse topology adopted throughout the exercise, refined distinctions could be meaningfully made.

In this application, we extend the arithmetic of spreads to an arithmetic of assessment, in simple sums of the sort encountered in tabulated statistical data. Here we will use the term grade, in the American sense of an index of quality, of the sort that are routinely assigned in innumerable spheres of activity in our society. (This differs from the British usage, as in "grading" of hotels or restaurants, mentioned above.) Let us first consider the rules of arithmetic which are appropriate for spread. These follow the traditional rules of the calculus of errors: in sums and differences, absolute errors add; while in multiplications and divisions, proportional errors add. We could adopt the common rule that the root-mean-square-sum of errors is to be taken, but this would introduce an inappropriate degree of complexity of calculation at this stage. Thus, (with a, b, c, d all positive)

$$(A + a) + (B + b) = (A + B) \pm (a + b)$$
$$(A \pm a) - (B \pm b) = (A - B) \pm (a + b)$$

and

$$(C \pm c\%) \times (D \pm d\%) = (C \times D) \pm (c + d)\%$$
$$(C \pm c\%)/(D \pm d\%) = (C/D) \pm (c + d)\%$$

The use of percentages to express proportional spreads must be done with caution, as it is very easy to write meaningless percentages. When spreads are as large as, or larger than the number itself, then the proper expression of proportional spread requires some skill. Quite reasonable percentages can give quite large spreads; thus spreads of $\pm 33\%$, $\pm 50\%$ and $\pm 67\%$ produce variations through log-ranges of 2, 3 and 5 respectively (since, for example, $(1 + 67\%)/(1 - 67\%) = 5$). For larger proportional spreads than those, we should use a notation for logarithmic range.

The rules of elementary arithmetic for the grade are nearly as simple, although there are two exceptional cases to be observed. In the case of addition and subtraction, we usually take the weighted mean of the separate grades of the numbers. This reflects the intuitive judgement that the quality of the result should be the average grade for the collection. The strengths or weaknesses of the separate elements are given their influence, proportional to the size of that element. Using the square brackets to denote the grade, we have:

$$E,[e] \pm F,[f] = (E \pm F),[(E \times e + F \times f)/(E + F)]$$

The two exceptional cases both apply when the two terms are nearly equal. The reason for an exceptional grade is easier to see in the case of subtraction. For if we have two terms that are nearly equal, as say 95 and 92, then any uncertainties in the initial terms will be magnified in their difference. This is easy to see in the case of spread; if each has a spread of $\pm$ 1, then their difference will be 3 $\pm$ 2. The proportional spread goes from about $\pm$ 1 % for each of initial terms, to $\pm$ 67 % for the difference; quite an enormous change. It is hard to imagine such a number being in any way as reliable as either of the initial terms. Hence we must construct a rule, inevitably somewhat arbitrary in its details, for reducing the grade of the difference element, when its spread is so dramatically increased. This will be a more general, simple and coarse version of the rules for distinguishing the means of statistical distributions. We divide the rule into three cases. There is no change in grade when the ratio of the difference between the terms to their average spread is greater than five. If that ratio is less than two, the grade is reduced by one-half. In between, the reduction is linear, bearing in mind that grades are expressed to the nearest single digit only.

The exceptional rule for addition is not quite so completely automatic in its operation; it comes into play when two quantities which are derived from independent procedures are averaged or compared in some other way. In this case, there is a qualitative judgement that if two such quantities are equal, or nearly so by some appropriate criterion, then this correspondence serves as a corroboration of them both. Even if neither of them can be checked directly against the empirical reality that it is intended to measure, the unlikelihood that both have come to the same mistaken estimate serves as positive evidence that they are both likely to be correct. For this case, we can apply the above rule in the other direction: when the ratio of difference to (average) spread varies between two and five, the grade is *increased* linearly by up to one-half.

For multiplication of numbers, the rule for grade is simple; for here we adopt a "weak link" principle: the grade of the product is the minimum of the grades of the factors. Thus,

$$G,[g] \times H,[h] = (G \times H),[\text{Min} (g.h)]$$

For this rule, there are no exceptional cases.

We notice that the grade almost always decreases in calculations, sometimes quite drastically. It could be that many computations which are up to now accepted as reasonable, and as providing meaningful outputs, would on such a grading system be judged as being of very low quality. In particular, matrix-inversion operations, involving the sums and differences of many-factored products, would be especially vulnerable. The fault, however, might not lie in the peculiarities of a grading system, but rather in a class of mathematical operations over which there has hitherto been very little effective quality-control.

13.2. AN EXAMPLE: THE VALUATION OF ECOSYSTEMS

To illustrate the usefulness of the proposed arithmetic of grades, we carry it through for the example case of ecosystem valuation. We use as well documented study of the economic value of wetlands in Louisiana (Farber and Costanza, 1987, Costanza et al., 1989) which employed a number of different models and methods to arrive at an estimate to the total value of the ecosystem. The results from the original study are reproduced in Table XII.

There are two overall methods whose results are presented. The "willingness to pay" (WTP) based method enumerates the various components of ecosystems value and derives an independent estimate for each one. These components are then added to yield the total value. For example, shrimp production value was estimated as $10.85/ac/yr, and storm protection value as $128.30/ac/yr. Option and existence value are known to be important components of the total but no direct estimate was made for this ecosystem. A second method, "energy analysis" (EA), uses the total solar energy captured by the ecosystem as an indicator for its economic value. It is more comprehensive (in that it does not require adding individually measured components to arrive at the total) but the connection between energy captured and economic value is controversial. Finally, the "present value" of the ecosystem is calculated using various discount rates, based on the assumption that the ecosystems provide a constant stream of benefits into the indefinite future. In this

TABLE XII
Summary of wetland value estimates in 1983 dollars

Method	Annual value per acre	Present value per acre at discount rate 8%	3%
WTP based			
Shrimp	10.85	136	362
Menhaden	5.80	73	193
Oyster	8.04	100	268
Blue Crab	0.67	8	22
Total commercial			
Fishery	25.37	317	846
Trapping	12.04	151	401
Recreation	3.07	46	181
Storm protection	128.30	1915	7549
Subtotal	168.78	2429	8977
Option and existence values	?	?	?
EA based			
GPP conversion	509–847	6,400–10,600	17,000–28,000
Best-estimate	169–509	2,429–6,400	8,977–17,000

TABLE XIII
Simplified research pedigree matrix

Code	Theoretical phase Quality of model	Empirical phase Quality of data	Social phase Degree of acceptance
4	Established theory	Experimental data	Total
3	Theoretical model	Historical/field data	High
2	Computational model	Calculated data	Medium
1	Statistical processing	Educated guesses	Low
0	Definitions	Uneducated guesses	None

case: present value = (annual value)/(discount rate). The appropriate discount rate to use in such a situation is, however highly uncertain.

The pedigree used in this study is a simplified version of the Research Pedigree Matrix, as given in Chapter 10 (Table XIII). Pedigree codings are based on an analysis of the individual models and methods used. For example, the shrimp production estimate was based on a theoretical model relating wetland area to shrimp catch, using Historical/Field Data from National Marine Fisheries shrimp catch statistics; and measured wetland area in a procedure (regression analysis) which has high but not total peer-acceptance. Finally the grade for each estimate is given based on the average codes in the pedigree $(3 + 3 + 3)/12 = 0.7$. Note that the grades are rounded to one digit.

Table XIV is a recasting of the results in Table XII into the NUSAP system. Here the numerical values are given only to the appropriate degree of precision, and the spreads on each number are shown, (using only 10 % increments except for 25 % and 75 %). Several quantities are calculated in Table XIV using the NUSAP arithmetic described above. The total commercial fishery value is the sum of four components. Its spread is the weighted average of the percentage spreads of the components

$$(1E1 \times 9.1 + 6E0 \times 0.2 + 8E0 \times 0.3 + 1E0 \times 0.4) / 2.5E1 = 0.2$$

Its grade is the weighted average of its component grades

$$(1E1 \times 0.7 + 6E0 \times 0.5 + 8E0 \times 0.6) / 2.5E1 = 0.6$$

An estimate for option and existence value is given based on studies for other areas, but as its spread and grade indicate, this application it is definitely an order-of-magnitude estimate. The total WTP based value reflects the quantitative importance of option and existence values and their relatively low quality. We end with a spread of $\pm$ 40 % and a grade of 0.3 for this estimate.

The EA based estimate yielded a very similar quantity estimate to the WTP based estimate; and this is taken as corroborating evidence, since the likelihood that this would occur by chance is small. The average of the two methods is therefore of higher grade than either of the inputs (0.6 versus 0.5

TABLE XIV

NUSAP representation for the elements of the wetland valuation study

Element	Numeral	Unit*	Spread	Assessment	Pedigree
WTP based estimates					
Shrimp	1 E1	$/a/yr	± 10%	0.7	(3,3,3)
Menhaden	6 E0	$/a/yr	± 20%	0.5	(2,2,2)
Oyster	8 E0	$/a/yr	± 30%	0.6	(2,3,2)
Blue Crab	1 E0	$/a/yr	± 40%	0.6	(3,2,3)
Total commercial					
Fishery	2.5 E1	$/a/yr	± 20%	0.6	
Trapping	1.2 E1	$/a/yr	± 30%	0.5	(2,2,2)
Recreation	3 E0	$/a/yr	± 10%	0.8	(3,4,3)
Storm protection	1.3 E2	$/a/yr	± 20%	0.6	(2,3,2)
Subtotal	1.7 E2	$/a/yr	± 20%	0.6	
Option and					
existence values	5 E2	$/a/yr	± 50%	0.2	(1,0,1)
Total WTP	7 E2	$/a/yr	± 40%	0.3	
EA based GPP					
conversion	7 E2	$/a/yr	± 25%	0.5	(3,2,1)
Average of					
two methods	7 E2	$/a/yr	± 30%	0.6	
Discount rate	5 E0	%	± 50%	0.4	(1,3,1)
Present value	15 E3	$/a/yr	± 80%	0.4	

* Unit should read $_{1983}$/a/yr

and 0.3) and we are left with a reasonably high quality estimate of the total annual value production.

$$7E2 : \$_{1983} : \pm 30\% : [0.6]$$

Converting this to present value significantly reduces the data quality, however, because of the severe uncertainty about the discount rate. The spread of the present values goes to ± 80% (a log-range of 9) and the grade goes down to 0.4.

The NUSAP representation of the series of calculations that went into the estimation of the value of wetlands offers a clear picture of the data quality. It also allows the uncertainty in the final estimate to be easily communicated; and its directs research to those areas most likely to improve the quality of that final estimate.

13.3. RISK INDICES: A NUSAP ANALYSIS*

The risks of nuclear power have been dramatically demonstrated by the Chernobyl accident. Here we use the NUSAP approach to calculate and express what can reliably be said about the risks of six energy technologies.

*The following sections reproduce research done with C.W. Hope.

The results of the exercise explain why debates about nuclear power and other energy sources frequently generate much heat but little light. They show the importance of expressing risk assessment in a notation which protects them from misuse (Hope and Funtowicz, 1989).

In an earlier study (Fischhoff *et al.*, 1984), Multi Attribute Utility Study (MAUT) was used to combine into a Risk Index the different elements that make up the ambiguous concepts of the "risk" of an energy technology. We reproduce here just enough of the calculation to show how the conclusion was reached. The elements, or attributes, that were incorporated in the Risk Index were as follows:

– fatalities amongst workers;
– fatalities amongst members of the public;
– illness and injury that did not lead directly to death;

and two elements representing the degree of concern engendered by the technology:

– one labelled "unknown", because new and mysterious technologies scored badly; and
– the other labelled "dread", because technologies associated with war and catastrophe also scored badly.

The technique was applied to six energy technologies, starting from the scores shown in Table XV for each technology on each attribute.

– Public Deaths: 0 means no death, 100 means 10 deaths per Gigawatt/year.
– Worker Deaths: 0 means no death, 100 means 10 deaths per Gigawatt/year.
– Morbidity: 0 means no injury, 100 means 60 days lost per Megawatt/year.
– Unknown: Hill climbing scores 0, DNA research scores 100.
– Dread Risk: Home appliances score 0, nuclear weapons score 100.

To form the Risk Index it was also necessary to assign weights to the different attributes. The weight on an attribute represents the importance allocated to decreasing the score on that attribute by one point. Four separate sets of

TABLE XV
The scores of the six energy technologies

Attribute	Coal	Hydro	Large-scale wind	Small-scale wind	Nuclear	Conser-vation
Public deaths	80	10	20	5	10	5
Worker deaths	30	20	10	30	5	10
Morbidity	20	20	40	50	10	40
Unknown	70	60	90	50	80	40
Dread	50	50	40	20	90	10

TABLE XVI
Four possible sets of weights

Attribute	A	B	C	D
Public deaths	0.33	0.40	0.20	0.08
Worker deaths	0.33	0.20	0.05	0.04
Morbidity	0.33	0.20	0.05	0.40
Unknown	0	0.10	0.30	0.24
Dread	0	0.10	0.40	0.24
Sum of weights	1	1	1	1

weights were designed, to illustrate four of the perspectives that members of society might adopt. These were labelled weights A to D, and are reproduced in Table XVI.

Given these scores and weights, it was computationally simple to calculate the risk from the different technologies. For example, the risk from coal using the A set of weights was $0.33 \times 80 + 0.33 \times 30 + 0.33 \times 20 = 42.9$. The complete set of risk indices obtained from the four sets of weights are displayed in Table XVII, rounded to the nearest integer.

The Risk Indices for coal, small scale wind and conservation varied relatively little across the sets of weights that were investigated, whilst those for hydropower, large scale wind and particularly, nuclear power varied much more. Consequently coal ranked consistently badly, whereas nuclear power varied from safest to most risky, depending on the set of weights adopted. These variations occurred despite the assumption of complete agreement about the magnitude of the individual components of risk from each technology, as expressed by the single set of attribute scores that were used. This showed that the controversy over the safety of nuclear power may have reflected disagreements about the weights to be attached to different categories of harm.

Although this calculation gave an intuitively appealing result, it did have an air of precision that was not warranted by the facts. In particular, the scores

TABLE XVII
The risk indices of the six technologies

Sets of weights	Coal	Hydro	Large-scale wind	Small-scale wind	Nuclear	Conser-vation
A	43	17	23	28	8	18
B	54	23	31	25	24	17
C	60	42	50	28	63	20
D	39	36	49	38	46	29

reproduced in Table XIV were simply point estimates derived from a reading of the relevant literature. There were considerable differences of opinion in these publications. For example, the extreme values for worker deaths for coal were 0.7 and 8 deaths per Gigawatt year of electricity generated. In the calculation, this diversity of estimates was collapsed in a single score of 30, equivalent to three worker deaths per Gwyr. One might ask how the results would change if the whole range of values from 0.7 to 8 were used.

Another worry was the use which might be made of the final results. Selecting a single set of weights from Table XVII would give a Risk Index value for each technology, apparently precise to plus or minus one point. The authors of the study were concerned not to produce another institutionalized "magic number", carved in stone and produced whenever debates about nuclear power degenerated into propaganda. Yet that is exactly what the numbers in the table could have become. The results contained no protection against such a misleading use.

As mentioned above, MAUT is the technique in which the different technologies were evaluated. The first task for a NUSAP translation is to describe the status of MAUT as an aid to decision-making. This can be done by assigning a Pedigree to MAUT, which will condition the pedigree of any results obtained by its use. At present the pedigree of MAUT, using the Research Pedigree Matrix (Tables IV, V), is best described as

$$P(MAUT) = (3,1,2,2)$$

MAUT as a method has the structure of a Theoretically-based Model because it is based on an axiomatic structure built from postulates of rational behaviour (Keeney and Raiffa, 1976). It cannot aspire to the highest code, that of an Established Theory, since there is no compelling body of evidence to demonstrate that people do in general behave in such a rational manner; if anything the opposite is true (Wright, 1984). On the other hand, it is more than a simple model to aid calculation, devoid of theoretical content. MAUT works with subjective data representing personally held degress of belief in the likelihood of events occurring, or the relative merit of different outcomes. These beliefs could be derived from experimental or historic data, but in practice MAUT comes into its own when such a data are absent, and the alternative to using MAUT is to abandon any attempt at formal analysis. In these conditions, the skillful user of the technique will use Educated Guesses as inputs to the calculation. Hence this is the data input entry.

The subjectivity of inputs precludes a high peer-acceptance of the results of most MAUT analyses, since other experts could give other educated guesses as inputs. Even a medium level of peer-acceptance can be expected only if the analyst carrying out the calculation has an established reputation in the field. There is no prevailing paradigm in the field worked by MAUT. Other techniques include: Cost-Benefit Analysis, Social Judgement Theory (Hammond et al., 1980, Analytic Hierarchies (Saaty, 1980) or Soft Systems Analysis (Checkland, 1981). All have their ardent advocates, giving a Colleague Consensus of Competing Schools.

Using the Reliability Index discussed previously, we find that MAUT scores 8/16, or 50 %. This immediately sounds a note of caution: Do not expect any results of a MAUT analysis to command universal approval! The justification for using MAUT lies in the belief that the pedigree-based evaluative score for any other technique in such a murky area of applied analysis would be no higher or probably even lower.

We can now proceed to apply NUSAP to the scores of the various energy technologies. We start with the pedigree. As we discussed above, the Risk Index of any technology is obtained by multiplying the score of the technology on each attribute, and adding over all attributes. There is potential for disagreement about the attributes chosen. For instance, some might argue that morbidity should also be split into public and worker, or, conversely, that a single attribute to describe deaths would be sufficient. This is one reason why the pedigree-based evaluative score of MAUT is not higher. The pedigree will also be affected by the ambiguity and vagueness in the naming and the operational definition of the attributes that we choose. For example, the data input to worker deaths of a particular technology could be anything from Historic/Field Data in the best case, to Uneducated Guesses in the worst. The measurement procedure will also affect the assignment of pedigree to the scores. This will be particularly true of the credibility of the organization that carries out the measurement.

In the case of coal and hydro deaths and morbidity, many studies exist which use Theoretically-based Models to project a wealth of Historic Data into the future (Bliss *et al.*, 1979, Baecher *et al.*, 1980). There is a Total Peer-Acceptance of the result of these studies, and All but Cranks would agree that the techniques used are sensible. This yields a pedigree of (3,3,4,4) for these scores. For large scale wind deaths and morbidity, and nuclear power deaths, the theoretical structure of probabilistic risk assessment combined with a modest amount of experience is just as strong as for coal and hydro

Attribute	Coal	Hydro	Large scale wind	Small scale wind	Nuclear	Conservation
Public deaths	XXXXX	XXXXX	/////	.....	/////	.....
Worker deaths	XXXXX	XXXXX	/////	.....	/////	.....
Morbidity	XXXXX	XXXXX	/////	.....	-----	.....
Dread	.....	.....	.....	.....	.....	.....
	.....	.....	.....	.....	.....	.....
Key:						
Pedigree-based ranking:		High	Good	Fair	Poor	None
		XXXXX	/////	-----	.....	
		⅄ 12/16	⅄ 9/16	⅄ 6/16	⅄ 3/16	⅄ 0/16

Fig. 14. The pedigree-based evaluative ranking of the attribute scores for each technology.

(Birkhofer, 1980). However the data input has only the status of Calculated Data from partial experiments extrapolated to full scale working systems. Peer-Acceptance of the results is high, since even critics agree that the calculations have been properly carried out, but the field is one of Competing Schools where criticism of the applicability of probabilistic risk assessment is strong (Fischhoff et al., 1981). A pedigree of (3,2,3,2) results.

Less attention has been paid to the morbidity from nuclear power, since the more dramatic deaths from catastrophic accidents have tended to claim the headlines and the research effort. Consequently the studies of morbidity have tended to be more sketchy, leading to less acceptance of the results by the community. A pedigree of (2,2,2,2) would be fair for this score. The scores for small scale wind and conservation, and the unknown and dread risk for all technologies, have a very low pedigree. The Statistical Processing technique of factor analysis is used to impose some order on Educated Guesses describing people's psychological reactions to the different technologies (Fischhoff et al., 1978). There is a medium level of Peer-Acceptance since the researchers involved have a good reputation in the field, but it is obvious that we are dealing with an Embryonic Field (Spangler, 1981). The pedigree is (1,1,2,1). The pedigree-based evaluative scores are shown graphically in Figure 14. The key to this figure associates the numerical codes with a descriptive ranking system.

As with the pedigree, our knowledge of the literature enables us to assign a spread and assessment to each of the attribute scores, to complete their NUSAP representation (Table XVIII). Each spread is represented in the notation as a "factor of n" (more familiar than $\Lambda2$ for log-range). Each assessment is of the form "a %", giving a subjective confidence interval of finding the true value within the range specified in the spread.

There are two different levels of confidence in the attribute scores, 80 % and 95 %. For the purposes of comparison, it would be helpful to have each score expressed with the same assessment entry. We can make a trade-off between spread and assessment; fixing a common level of confidence a 80 %, we obtain

TABLE XVIII
The spread and assessment of the attribute scores for each technology

Attribute	Coal	Hydro	Large-scale wind	Small-scale wind	Nuclear	Conservation
Public deaths	f2,95%	f2,95%	f3,80%	f3,80%	f3,80%	f2,95%
Worker deaths	f2,95%	f2,95%	f3,80%	f3,80%	f3,80%	f2,95%
Morbidity	f2,95%	f2,95%	f3,80%	f3,80%	f3,80%	f2,95%
Unknown	f2,80%	f2,80%	f3,80%	f3,80%	f2,80%	f3,80%
Dread	f2,80%	f2,80%	f3,80%	f3,80%	f2,80%	f3,80%

a reduced spread entry of f 1.5. We now have a full NUSAP representation of each of the attribute scores that are the inputs to the Risk Index calculation. Those for coal are shown in Table XIX, as an illustration. The richness of information conveyed in the table emphasizes just how much was missing from ostensibly the "same" information conveyed by the bald and unqualified scores in the first column of Table XV.

The other inputs to the Risk Index are the weights representing the strength of feeling about each attribute. The four sets of weights shown in Table XVI were used in the original study, to give some idea of the range that might be encountered across different groups in society. The value that any individual weight can assume is circumscribed by the underlying theory of MAUT, which states that:

- Since each weight represents an individual's strength of feeling, it can take only a single value, not a range or a probability distribution. This is the same restriction that precludes the use of probabilities of probabilities.
- The weights must add to one across all the attributes. This is necessary if comparison between different sets of weights is to be meaningful.

Each individual weight will therefore have numeral, unit and spread entries of $w : 1 - $, where w is the value of the weight as shown in the table, and " $ - $ " in spread functions as a filler, since there is no spread. Leaving assessment to one side for a moment, we ask what the pedigree of the weight will be. The theoretical structure of MAUT, which we have already discussed is a Theoretically-based Model. The Data Input is a subjective Educated Guess; despite the best endeavours of a generation of economists, there is as yet no compelling evidence that any one set of weights is more reasonable than any other (Pearce, 1979, Hope and Owens, 1986). Because the weights are a matter of subjective judgement, the Peer-Acceptance of any set will be low, and the state of the field is characterized by Competing Schools, with some authors supporting "objective" market forces or community valuation, and others favouring "subjective" individual introspection (Keeney, 1977), or decision conferencing (Phillips, 1984), to arrive at a reasonable weight values. These considerations give a pedigree for each set of weights of

$$P(\text{weights}) = (3,1,1,2)$$

Finally for weights, the assessment category must be filled. We cannot have anything like the 80 % or 95 % confidence intervals that we used for scores, since each weight is only allowed to take a single value. Instead, on the absence of anything better for assessment, we can use the pedigree-based evaluative ranking as a Reliability Index. For the weights it is 7 out of 16, hence their NUSAP representations is:

$$w : 1 : - : 7/16 : (3,1,1,2)$$

13.4. CALCULATING THE RISK INDICES FOR ENERGY TECHNOLOGIES

We are now in a position to start the calculation, having expressed each of our inputs in the NUSAP notation. Only the calculation for coal is displayed in full here; the calculation for the other technologies follows an identical path. The Risk Index for coal on each set of weights will be the sum of five terms (one for, each attribute), each of which is the score of the technology on that attribute times the weight assigned to the attribute. Each score is as described in Table XIX in the NUSAP scheme.

The notation for the spread category, a factor of 1.5 or 2, tells us that a log-normal probability distribution would be an appropriate description for each score, rather than a normal distribution which would have been appropriate if the spread had been expressed as, say $\pm 1/10$. Since the weights are single values, each term of score $\times$ weight will also have a log-normal distribution. However, the central limit theorem tells us that the sum of the terms will tend towards a normal distribution, provided the individual terms are independent. Even with only five terms, as here, the distribution of the Risk Index will be close to normal. The only mathematics required is to calculate the mean and standard deviation of the sum of the terms.

The mean of the sum of terms is just equal to the sum of the means of the individual terms. Using the B set of weights from Table XVI as an illustration, the mean of the Risk Index for coal will be

$$80 \times 0.4 + 30 \times 0.4 + 20 \times 0.2 + 70 \times 0.1 + 50 \times 0.1 = 54$$

This is the same value as shown in table XVII.

The standard deviation of the sum of terms is equal to the square root of the sum of the squared standard deviations of the individual terms. The standard deviation, S, of each score can be found by recalling that for log-normal distributions there is an 80 % chance that the score will be found in a credible span of length about 2.5S (and a 95 % chance of being in a credible span of 4S), from a lower limit, L, to an upper limit, U, located such that U/M = M/L. So S = (U − L)/2.5 for an 80 % cre `ble span; S = (U − L)/4 for a 95 % credible span (R.L. Brown, 1971). Taking Public Deaths as an

TABLE XIX
NUSAP representation of the attribute scores for coal

Attribute	N		U		S		A		P
Public deaths	8	:	10	:	f1.5	:	80%	:	(3,3,4,4)
Worker deaths	3	:	10	:	f1.5	:	80%	:	(3,3,4,4)
Morbidity	2	:	10	:	f1.5	:	80%	:	(3,3,4,4)
Unknown	7	:	10	:	f2	:	80%	:	(1,1,2,1)
Dread	5	:	₁0	:	f2	:	80%	:	(1,1,2,1)

example, the score has a 10 % chance of being more than 1.5 times M, and a 10 % chance of being less than 1/1.5 times M, where M = 80. So U = 120, L = 53 and S = 27. Similarly, the standard deviation for Worker Deaths is 10, for Morbidity 7, for Unknown Risk 47, and for Dread Risk 34.

Using the B set of weights for illustration once again, we can calculate the standard deviation of the Risk Index as

$$\sqrt{[(0.4 \times 27)^2 + (0.2 \times 10)^2 + (0.2 \times 7)^2 + (0.1 \times 47)^2 + (0.1 \times 34)^2]}$$

which gives a standard deviation of 12.5.

The calculation complete, we have only to recall that 80 % of a normal distribution lies within 1.28 standard deviations of the mean to deduce that the calculated value of the Risk Index for coal on the B set of weights has an 80 % chance of lying within $1.28 \times 12.5 = 16$ points of the central value of 54.

How should this result be described in the NUSAP notation? It would be possible to write it as

$$54 : 1 : \pm 16 : 80\% : (a,b,c,d)$$

Guiven the uncertainties and approximations implicit in the data and calculations, however, there is an obvious pseudo-precision that is immediately detectable from the number itself. If its spread is 16, how can we possible be confident that its central value is 54, rather than 53 or 55? We cannot; therefore one correct treatment of the result would be to spread the number a little more, firm up the assessment slightly, and write

$$5 : 10 : \pm 2 : 90\% : (a,b,c,d)$$

Here the spread and numeral refer to the same unit.

The pedigree of this result is affected by the pedigrees of each of the components used to obtain it. Each of the five attribute scores, and each of the five weights. One simple way of compounding pedigrees is to take the average of their Reliability Index. The values for the five scores are 14, 14, 14, 5, 5 and for the five weights 7, 7, 7, 7, 7. The average of these values is 8.5, which translates to a verbal reliability of Fair, using the key to Figure XIV as our guide. we now have our first complete result:

$$\text{Risk Index (Coal, B)} = 5 : 10 : \pm 2 : 90\% : \text{Fair}$$

The results for the other technologies, and coal using the other sets of weights, follow in an analogous manner, to give the means and standard deviations of the Risk Indices shown in Table XX.

For some technologies, on some sets of weights, the standard deviations are as large as the mean values. This already gives some feeling for the imprecision of estimates of risk given the present state of knowledge. This can be rein-forced by completing the NUSAP representation, as was done for coal on the

TABLE XX
Means and standard deviations of the risk indices

Sets of weights	Coal	Hydro	Large-scale wind	Small-scale wind	Nuclear	Conservation
A	43 (10)	17 (4)	23 (23)	28 (30)	8 (8)	18 (5)
B	54 (13)	23 (6)	31 (23)	25 (20)	24 (11)	17 (7)
C	60 (20)	42 (18)	50 (47)	28 (26)	63 (29)	20 (19)
D	39 (13)	36 (13)	49 (43)	38 (35)	46 (20)	29 (15)

Note: Each entry is mean (standard deviation)

B set of weights above. Thus, the Risk Indices for coal, for the four sets of weights, are:

$$\text{Risk Index (Coal, A)} = 4:10:\ =\ :90\%:\text{Fair}$$
$$\text{Risk Index (Coal, B)} = 5:10:\pm 2:90\%:\text{Fair}$$
$$\text{Risk Index (Coal, C)} = 6:10:\pm 3:90\%:\text{Fair}$$
$$\text{Risk Index (Coal, D)} = 4:10:\pm 2:90\%:\text{Fair}$$

All the representations look similar to the calculation that we followed in detail, for the B set of weights. The only slight difference in notation is for the A set of weights, where there is a " = " in the spread position. This is a shorthand way of recording that the spread is the same as the unit. With the assessment of 90 %, it implies that there is only a 10 % chance that the Risk Index lies outside the range adequately described by the numeral and unit together: A Risk Index of four tens.

The Risk Indices for Hydro are as follows:

$$\text{Risk Index (Hydro, A)} = 17:1:\pm 25\%:90\%:\text{Fair}$$
$$\text{Risk Index (Hydro, B)} = 2:10:\ =\ :90\%:\text{Fair}$$
$$\text{Risk Index (Hydro, C)} = 4:10:\pm 2:90\%:\text{Fair}$$
$$\text{Risk Index (Hydro, D)} = 4:10:\pm 2:90\%:\text{Fair}$$

These expressions are of a similar form of those for coal, with the exception of the A set of weights. The uncertainties in the index are small enough that meaningful information would be lost if the result were expressed to the nearest ten. The spread is also more intuitively appealing if written as a percentage range: it is small enough here for such an approximation to be permissable. In the full NUSAP notation, the Hydro Risk Indices for the C and D sets of weights are indistinguishable. The apparent differences for the mean and standard deviation shown in Table XIX are revealed as pseudo-precision when the full range of qualifying factors is taken into account.

The indices for Large Scale Windpower require some new notations. On the set A of weights, "U" in the assessment entry stands for an Upper Limit indicator. For this set of weights, the Risk Index of Large Scale Windpower is unlikely to be greater than five tens. The qualifications around this result are so great that "unlikely" is as precise an expression as is warranted; "a 5 % chance" would be pseudo-precision. With the B set of weights, even this degree of precision is not appropriate, the " ~ " for the spread and assessment entries tell us that all we have here is an order-of-magnitude estimate. The 3 in the numeral position gives notice that the Risk Index is more likely to lie towards the (geometric) centre of the order-of-magnitude than at the extremes of one ten or ten tens. With the C and D sets of weights, even this is removed. The " – " in the numeral position is a filler as before, indicating that we have no meaningful information about where in the order-of-magnitude range the Risk-Index is likely to lie. All that we can say is that it will be of order of ten, rather than one or a hundred. The four NUSAP representations are as follows:

$$Risk\ Index\ (Lsw, A) = 5 : 10 : - : U : Fair$$
$$Risk\ Index\ (Lsw, B) = 3 : 10 : \sim : \sim : Fair$$
$$Risk\ Index\ (Lsw, C) = - : 10 : \sim : \sim : Fair$$
$$Risk\ Index\ (Lsw, D) = - : 10 : \sim : \sim : Fair$$

The representations of the indices for Small Scale Windpower are similar to those for its large scale cousin. If anything the reliability is even lower here, as reflected by the "poor" entry in pedigree.

$$Risk\ Index\ (Ssw, A) = 3 : 10 : \sim : \sim : Poor$$
$$Risk\ Index\ (Ssw, B) = 5 : 10 : - : U : Poor$$
$$Risk\ Index\ (Ssw, C) = 3 : 10 : \sim : \sim : Poor$$
$$Risk\ Index\ (Ssw, D) = - : 10 : \sim : \sim : Poor$$

In the original study, nuclear power had the distinction that the value of its Risk Index showed the greatest variation of all the technologies. The NUSAP expressions for Nuclear are similarly the most varied in notational representation. If the B set of weights is believed, the risk from Nuclear power is low and quite precisely known, with a spread of only about one ten. On the set A an even more reassuring picture emerges. The $\%$ 90 assessment entry is, as we have seen before, a shorthand for the 90th percentile of the probability distribution: there is only a 10 % chance that nuclear power has a worst risk than this. The D set of weights gives an index that is spread more, but can still be expressed as a confidence interval. On the other hand, the C set give an index of the most imprecise nature, of the order of ten rather than one or a hundred is all that can be said.

$$Risk\ Index\ (Nuc, A) = 2 : 10 : - : 90\% : Fair$$
$$Risk\ Index\ (Nuc, B) = 2 : 10 : - : 90\% : Fair$$
$$Risk\ Index\ (Nuc, C) = - : 10 : \sim : \sim : Fair$$
$$Risk\ Index\ (Nuc, D) = 5 : 10 : \pm 3 : 90\% : Fair$$

For Conservation, the indices are uniformly low, but once again of varying precision. The A and B weights yield fairly tight confidence intervals, while

C and D give orders-of-magnitude centre around the low tens. The reliability of the Conservation results is low because of the diversity of technologies that go to make up Conservation, and the lack of studies devoted to gauge their risk. The NUSAP representation for Nuclear and Conservation are as follows:

$$\text{Risk Index (Cons, A)} = 2:10:-:90\%:\text{Poor}$$
$$\text{Risk Index (Cons, B)} = 2:10:-:90\%:\text{Poor}$$
$$\text{Risk Index (Cons, C)} = 2:10:\sim:\ \sim\ :\text{Poor}$$
$$\text{Risk Index (Cons, D)} = 3:10:\sim:\ \sim\ :\text{Poor}$$

The flexibility of the NUSAP system allows us to attempt to go one stage further here than was possible in the original study, by aggregating the results for each technology across the four sets of weights. We recall that each set of weights has been chosen to characterize one particular view that might be held inside society about the relative importance of a worker death against that of a member of the public, and so on. Each set of weights is merely the outward manifestation of a complex value system as is relates to this particular small problem. In the absence of obvious internal incoherence, it is not possible to dismiss any of the sets of weights as false, wrong or irrelevant. Thus it is not sufficient to choose whichever of the sets of weights takes our fancy and say that the results for this set are the ones that society should use in deciding which energy technologies to adopt. There does not exist a reliable empirical sociological study to tell us how the different sets of weight values poll amongst members of the public (Prescott-Clarke, 1982), so we cannot even form some sort of average index. All that we can do is use the results for our characteristic sets of weights to form an envelope inside which the views of society can be presumed to lie. This yields the following Risk Indices:

$$\text{Risk Index (Coal)} \quad = 1/2:100:\pm 1/3:90\%:\text{Fair}$$
$$\text{Risk Index (Hydro)} \ = 1/3:100:\ \text{f2}\ :90\%:\text{Fair}$$
$$\text{Risk Index (Lsw)} \quad = -\ :\ 10:\ \sim\ :\ \sim\ :\text{Fair}$$
$$\text{Risk Index (Ssw)} \quad = 1\ :100:\ -\ :\ \text{U}\ :\text{Poor}$$
$$\text{Risk Index (Nuclear)} = -\ :\ 10:\ \sim\ :\ \sim\ :\text{Fair}$$
$$\text{Risk Index (Cons)} \quad = 1/2:100:\ -\ :\ \text{U}\ :\text{Poor}$$

Now we can see why debates about risk often yield so much heat and so little light. When all the uncertainties and disagreements about value structures are taken into account, there is no definitive answer that quantitative techniques can give at present about the relative riskiness of different energy technologies. The NUSAP representations of the Risk Indices overlap to such a large degree that there is not a single ranking of technologies from the best to worst that we can consider to be "disproved by the facts". We interpret the results to say that society should probably view Hydropower and Conservation as less risky than Coal or Small Scale Windpower, with Nuclear power and Large Scale Windpower not locatable on the scale between them. However, if anyone wishes to interpret the results to argue that society should be more wary of Hydro and Conservation than all the other technologies, we would have to agree that such an interpretation is possible.

EPILOGUE

With the applications of NUSAP in our latter chapters we have shown how an apparently simple notational scheme can be used as a framework for sophisticated reasoning about uncertainty. In skilled hands NUSAP can be effectively applied wherever uncertainty is an issue. Once there is an identification of the elements of the uncertainty problem (including the actors along with cognitive and pragmatic aspects) an appropriate form of NUSAP can be devised. In policy-related research the cognitive uncertainties, frequently severe and now politicized as well, can be translated through NUSAP into evaluation of quality of policy inputs.

The effectiveness of NUSAP depends on its coherence; this is provided by a particular philosophical understanding and historical perspective. Based on these, we identify with, and participate in, the new shared awareness of our time. This concerns the destructive impact of our inherited high-technology civilization on the natural environment; and we understand this as the leading contradiction for humanity in our time. The complexity of the issues arising from that contradiction are forcing a new conception of knowledge, involving a recognition of its dialectical character. We can no longer sustain the traditional oppositions, of knowing-that and knowing-how, of formal and informal knowledge, of facts and values and, most basically, of knowledge and ignorance. It depends on our common awareness and skills, whether these complementary pairs are synthesized as creative contradictions leading to new understanding; or whether they remain stagnant and destructive oppositions.

The possibility of the diffusion of the elements of this philosophical conception, so necessary for our survival, represents the positive side of that leading contradiction. For the high-technology system not only provides opportunities for literate culture for growing sections of the population; it also needs a broadly diffused high level of literacy for its effective operation. Hence politics changes its form; the old oppositions between a rational and sophisticated elite minority and an ignorant, oppressed majority, are being transformed. Ideals of quality of life that transcend the never-ending accumulation of material objects are now politically effective. The new politics of "participation" requires a broad sharing of knowledge; and therefore the skills and power.

Numbers, however, are still esoteric knowledge, the property of a small set of initiates who control their magic power on behalf of their masters. This magic has continued to dominate policy decisions and debates. Unlike literacy, numeracy is still not effectively taught. This failure reflects the continued dominance of an obsolete philosophy of mathematics and of knowledge. But just as the operation of a high-technology society requires a broadly diffused literacy, so the struggle to resolve its contradictions requires a broadly diffused numeracy. Only when there is effective quality control of science for policy, through the management of uncertainties, will we be able to cope intelligently with the crises we face. The demystification of the mathematics of uncertainty is therefore a central part of the programme for the democratization of scientific expertise. This is our contribution.

REFERENCES

Agarwal, A. *et al.*: 1981, *Water, Sanitation, Health, Space for All?*, IIED, London.

Akehurst, R.: 1986, *Planning Hospital Services – An Option Appraisal of a Major Health Service Rationalisation*, Centre for Health Economics, Discussion Paper 12, York.

Appel, K. and W. Haken: 1978, The Four Color Problem. In *Mathematics Today*, L. A. Steen (ed.), Springer Verlag, New York, pp. 153–190.

Bachelard, G.: 1938, *La formation de l'esprit Scientifique*, Vrin, Paris.

Bacon, F.: 1621, *Novum Organum*, **1**, Aphorism 98.

Baecher, G. B., Pate, M. E. and de Neufville, R.: 1980, Risk of Dam Failure in Benefit-Cost Analysis, *Water Resources Research*, **16**, 449–456.

Bagrow, L.: 1964, *A History of Cartography*, Watts, London.

Bailar, J. C.: 1988, *Scientific Inferences and Environmental Problems: The Uses of Statistical Thinking*, Institute for Environmental Studies, The University of North Carolina, Chapel Hill.

Beck, M. B.: 1987, Water Quality Modelling: A Review of the Analysis of Uncertainty, *Water Resources Research*, **38**(8), 1393–1442.

Birkhofer, A.: 1980, The German Risk Study for Nuclear Power Plants, *IAEA Bulletin*, **22**, 23–33.

Bliss, C. *et al.*: 1979, Accidents and Unscheduled Events Associated with Non-Nuclear Energy Resources and Technology, MITRE Corporation, for the Environmental Protection Agency, Washington D.C.

Bridgman, P. W.: 1927, *The Logic of Modern Physics*, Macmillan, London.

Bridgman, P. W.: 1931, *Dimensional Analysis*, Yale U.P., New Haven.

British Standards Institution: 1975, *BS 2846: Part 1*, London, pp. 22–23.

British Standards Institution: 1979, *BS 4778*, London.

Brooks, H. and A. MacDonald: 1988, Letters, *Science*, **242**, 496.

Brown, H. S.: 1987, *A Critical Review of Current Approaches to Determining "How Clean Is Clean" at Hazardous Waste Sites*, CENTED Reprint 58, Clark University, Worcester (MA).

Brown, R. L.: 1971, *Research and the Credibility of Estimates*, Irwin, Homewood (Ill).

Brownstein, L.: 1987, Relevance of the Rationalist Constructivist Relativistic Controversy for the Validation of Scientific Knowledge Claims, *Knowledge*, **9**, 1, 117–144.

Cajori, F.: 1928, *A History of Mathematical Notations*, The Open Court, La Salle (Ill.).

Campbell, N. R.: 1920, *Physics: The Elements*, Cambridge U.P.

Campbell, N. R.: 1921, *What Is Science?*, Methuen, London.

Campbell, N. R.: 1928, *An Account of the Principle of Measurement and Calculation*, Longmans, Green, London.

Cantley, M.: 1987, Democracy and Biotechnology: Popular Attitudes, Information, Trust and the Public Interest, *Swiss Biotech*, **5**, 5–15. Quoted in Otway, 1988.

Carroll, L.: 1893, *Sylvie and Bruno Concluded*. In *The Complete Works of Lewis Carroll*, Random House, New York, and Nonsuch Press, London (1939).

Checkland, P.: 1981, *Systems Thinking, Systems Practice*, Wiley, Chichester.

Clark, W. C. and G. Majone: 1985, The Critical Appraisal of Scientific Inquiries with Policy Inplications. In *Science, Technology and Human Values*, **10**(3), 6–19.

Costanza, R. and H. Daly: 1987, Towards an Ecological Economics, *Ecological Modelling*, **38**, 1–12.

Costanza, R., S. C. Faber and J. Maxwell: 1989, The Valuation and Management of Wetland Ecosystems, *Ecological Economics* (forthcoming).

Covello, V. T.: 1987, Case Studies of Risk Communication: Introduction. In *Risk Communication*, Davis, F. *et al.* (eds.), The Conservation Foundation, Washington D.C., pp. 63–65. Quoted in Otway, 1988.

Crosland, M. P.: 1962, *Historical Studies in the Language of Chemistry*, Heinemann.

Dalkey, N.: 1969, An Experimental Study of Group Opinion. The Delphi Method, *Futures*, **1**(5), 408–426. Quoted in Mosteller, 1977.

Dasgupta, A. K. and D. Pearce: 1972, *Cost-Benefit Analysis*, Macmillan.

Davis, P. J. and R. Hersh: 1981, *The Mathematical Experience*, The Harvester Press, Brighton (Sussex).

Dieudonne, J.: 1970, The work of Nicholas Bourbaki, *American Mathematical Monthly*, **77**, 134–145.

Doebelin, E. D.: 1986, *Measurement Systems: Applications and Design*, McGraw-Hill.

Douglas, M. and D. Wildavski: 1982, *Risk and Culture*, University of California.

Duhem, P. M. M.: 1914, *La théory physique. Son object – sa structure*, Bibliothèque de Philosophie Expérimentale, Riviere, Paris.

Dummet, M.: 1977, *Elements of Intuitionism*, Clarendon Press, Oxford.

Dunster, H. J. and M. Vinck: 1979, The Assessment of Risk – Its Value and Limitations. In *European Nuclear Conference – Foratom VII Congress*, Hamburg, Essen, Vulkan-Verlag, pp. 162–166. Quoted in The Royal Society, 1983, 38.

Eco, U.: 1984, *The Name of the Rose*, Pan Books, London.

Eisenbud, M.: 1987, *Environmental Radioactivity*, Academic Press, New York.

Ellis, B.: 1966, *Basic Concept of Measurement*, Cambridge U.P.

Escher, M. C. *et al.*: 1972, *The World of M. C. Escher*, H. N. Abrams, New York.

Farber, S. and R. Costanza: 1987, The Economic Value of Wetlands Systems, *J. of Environmental Management*, **24**, 41–51.

Fairley, W. B.: 1977, Evaluating 'Small' Probability of a Catastrophic Accident from the Marine Transportation of Liquified-Natural Gas. In *Statistics and Public Policy*, Fairley, W. B. and F. Mosteller (eds.), Addison-Wesley.

Feinstein, A.: 1977, *Clinical Biostatistics*, Mosby, St. Louis.

Feyerabend, P. K.: 1975, *Against Method*, New Left Books, London.

Fischhoff, B., Slovic, P., Lichtenstein, S., Read, S. and B. Combs: 1978, How Safe Is Safe Enough?, *Policy Sciences*, **8**, 127–152.

Fischhoff, B., Lichtenstein, S., Slovic, P., Derby, S. L. and R. L. Keeney: 1981, *Acceptable Risk*, Cambridge U.P., New York.

Fischhoff, B., Watson, S. R. and C. W. Hope: 1984, Defining Risk, *Policy Sciences*, **17**, 123–139.

Fox, R.: 1974, The Rise and Fall of Laplacian Physics, *Historical Studies in the Physical Sciences*, **4**, 89–136.

Funtowicz, S. O. and J. R. Ravetz: 1984, Uncertainties and Ignorance in Policy Analysis, *Risk Analysis*, **4**, 219–220.

Funtowicz, S. O. and J. R. Ravetz: 1985, Three Types of Risk Assessment. A Methodological Analysis. In *Risk Analysis in the Private Sector*, Whipple, C. and V. T. Covello (eds.), Plenum, New York, pp. 217–232.

Funtowicz, S. O. and J. R. Ravetz: 1986, Policy Related Research: A Notational Scheme for the Expression of Quantitative Technical Information, *J. Opl. Res. Soc.*, **37**(3), 1–5.

Funtowicz, S. O. and J. R. Ravetz: 1987a, Qualified Quantities – Towards an Arithmetic of Real Experience. In *Measurement, Realism and Objectivity*, J. Forge (ed.), Dordrecht, Reidel, 59–88.

Funtowicz, S. O. and J. R. Ravetz: 1987b, The Arithmetic of Scientific Uncertainty, *Phys. Bull.*, **38**, 412–414.

Funtowicz, S. O. and J. R. Ravetz: 1989, Managing the Uncertainties of Statistical Information. In *Environmental Threats: Social Science Studies in Risk Perception and Risk Management*, J. Brown (ed.), Pinter, London.

Funtowicz, S. O., Macgill, S. M. and J. R. Ravetz: 1988, Mapping Uncertainties of Radiological Hazards, *Atom*, November, 15–16.

Funtowicz, S. O., Macgill, S. M. and J. R. Ravetz: 1989, The Management of Uncertainties in Radiological Data, *J. Radiol. Prot.* **9**, 257–261.

Funtowicz, S. O., Macgill, S.M. and J.R. Ravetz: 1989, Quality Assessment of Radiological Model Parameters, *J. Radiol. Prot.* **9**, 263–270.

Funtowicz, S.O., Macgill, S.M. and J.R. Ravetz: 1989, The Propagation of Parameter Uncertainties in Radiological Assessment Models, *J. Radiol. Prot.* **9**. 271–280.

Galilei, G.: 1632, *Dialogue Concerning the Two Chief World Systems*, University of California (1953).

Gavaghan, H.: 1989, Computers in Cockpits Breed Pilot Complacency, *New Scientist*, 26 August, 33.

Gherardi, S. and B. Turner: 1987, *Real Men don't Collect Soft Data*, Quaderno 13, D. di Politica Sociale, Universita di Trento.

Gleik, J.: 1988, *Chaos*, Heinemann, London.

Goguen, J.: 1979, Fuzzy Sets and the Social Nature of Truth. In *Advances in Fuzzy Set Theory and Applications*, M. M. Gupta *et al.* (eds.), North-Holland, Amsterdam, pp. 49–67.

Hall, M. C. G.: 1985, Estimating the Reliability of Climate Model Projections – Steps towards a Solution. In *The Potential Climate Effects of Increasing Carbon Dioxide*, MacCracken, M. C. and F. M. Luther (eds.), US Department of Energy, DOE/ER-0237, Washington D. C., 337–364.

Hammond, K. R., Rohrbaugh, J., Mumpower, J. and J. Adelman: 1980, Social Judgement Theory: Applications in Policy Formation. In *Human Judgement and Decision Processes in Applied Settings*, Kaplan M. F. and S. Schwartz (eds.), Academic press, New York.

Hazardous Waste Inspectorate: 1985, *First Report – Hazardous Waste Management: An Overview*, Department of Environment, London.

Henrion, M. and B. Fischhoff: 1986, Assessing Uncertainties in Physical Constants, *Am. J. Phys.*, **54**(9), 791–798.

Henrion, M.: 1988, Uncertain Information Processing in Knowledge Support Systems. In *Concise Encyclopedia of Information Processing Systems and Organizations*, A. P. Sage (ed.), Pergamon Press, Oxford.

Hogben, L.: 1968, *Mathematics for the Million*, W.W.Norton, New York.

Hope, C. W. and P. H. Gaskell: 1985, The Competitive Price of Oil: Some Results under Uncertainty, *Energy Economics*, 289–296.

Hope, C. W. and S. Owens: 1986, Frameworks for Studying Energy and the Environment, *Environment and Planning A,* **18**, 851–864.

Hope, C. W. and S. O. Funtowicz: 1989, Describing Risk. In *Proceedings of ENVRISK'88*, Como, Italy (forthcoming).

Hospers, J.: 1967, *An Introduction to Philosophical Analysis*, Routledge and Kegan Paul, London.

Hotelling, H.: 1931, The Economics of Exhaustible Resources, *J. of Political Economy*, **39**, 137–175.

Huff, D.: 1954, *How to Lie with Statistics*, Gollancz, London.

Hughes, J. A.: 1989: Personal Communication.

Irvine, J. *et al.*: 1979, *Demystifying Social Statistics*, Pluto Press, London.

Jaffe, A. J. and N. F. Spirer: 1987, *Misused Statistics*, Dekker, New York.

Keeney, R. L.: 1977, The Art of Assessing Multiattribute Utility Functions, *Organizational Behaviour and Human Performance*, **19**, 267–310.

Keeney, R. L. and H. Raiffa: 1976, *Decisions with Multiple Objectives: Preferences and Value Tradeoffs*, Wiley, New York.

Keyfitz, N.: 1988, Letters, *Science*, **242**, 496.

Keynes, J. M.: 1921, *A Treatise on Probability*, St. Martin's Press (1952), New York.

Kidron, M. and R. Segal: 1981, *The State of the World Atlas*, Heinemann, London.

Kleene, S. C.: 1964, *Introduction to Metamathematics*, North-Holland, Amsterdam.

Kline, M.: 1974, *Why Johnny Can't Add: The failure of the New Math*, Vintage Books, New York.

Kline, M.: 1980, *Mathematics: The Loss of Certainty*, Oxford U.P., New York.

Kripke, S. A.: 1982, *Wittgenstein on Rules and Private Language*, Blackwell, Oxford.

Kuhn, T. S.: 1961, The Function of Measurement in Modern Physical Science, *Isis*, **LV**, 161–193.

Kuhn, T. S.: 1962, *The Structure of Scientific Revolutions*, University of Chicago.

Lakatos, I.: 1976, *Proofs and Refutations*, Cambridge U.P.

Lester, S. U.: 1989, The Use of Science in Government – Don't Bother Me With The Facts, *Everyone's Back Yard*, Citizens's Clearinghouse for Hazardous Wastes Inc., **7**(1), 7–8.

Lewis, H. W. *et al.*: 1978, *Risk Assessment Review Group*. Report to the U.S. Nuclear Regulatory Commission, NUREG/CR-0400, Washington D.C.

Lucas, J. R.: 1970, *The Concept of Probability*, Clarendon Press, Oxford.

Mach, E.: 1883, *The Science of Mechanics*, The Open Court, La Salle (Ill.) (1942).

Macgill, S. M. and S. O. Funtowicz: 1988, The 'Pedigree' of Radiation Estimates, *J. Radiol. Prot.*, **8**, 77–86.

Mackay, A. L.: 1977, *The Harvest of a Quiet Eye*, The Institute of Physics, Bristol and London.

Mac Lane, S.: 1988, Letters, *Science*, **241**, 1144, and **242**, 1623–1624.

Maddox, J.: 1987, Half-Truths Make Sense (Almost), *Nature*, **236**, 637.

Majone, G.: 1989, *Evidence, Argument and Persuasion in the Policy Process*, Yale U.P., New Haven.

Moravscik, M. J.: 1985, Applied Scientometrics: An Assessment Methodology for Developing Countries, *Scientometrics*, **7**, 165–176.

Mosteller, F.: 1977, Assessing Unknown Numbers: Order of Magnitude Estimation. In *Statistics and Public Policy*, Fairley, W. B. and F. Mosteller (eds.), Addison-Wesley, pp. 163–184.

National Cancer Institute: 1981, Carcinogenesis Bioassay of 1, 2 Dibromoethane (Inhalation Study) TR-210 (CAS N.106-93-4), *Carcinogenesis Testing Program DHSS Publ.N.(NIH)*, 81–1766. Quoted in Whittemore, 1983.

National Radiological Protection Board (NRPB): 1986, Private Communication.

National Radiological Protection Board (NRPB): 1987, Private Communication.

Needham, J.: 1956, *Science and Civilization in China*, Cambridge U.P.

Nelkin, D. (ed.): 1979, *Controversy: Politics of Technical Decisions*, Sage Publications, London.

Olby, R. C.: 1966, *Origins of Mendelism*, Constable, London, 116/182–185.

Otway, H.: 1988, *Reflections on Communication about High-Risk Technologies*, Technical Note I.88.114, PER 1578/88, Joint Research Centre, Ispra Establishment.

Pacey, A. (ed.): 1977, *Water for the Thousand Millions*, Pergamon, Oxford.

Pacey, A.: 1983, *The Culture of Technology*, Blackwell, Oxford.

Page, T.: 1978, A Generic View of Toxic Chemicals and Similar Risks, *Ecological Law Quaterly*, **7**, 207–244. Quoted in Whittemore, 1983.

Pearce, D. W.: 1979, Social Cost-Benefit Analysis and Nuclear Futures, *Energy Economics*, **1**, 66–71.

Pearson, K.: 1892, *The Grammar of Science*, Everyman, London (1937).

Phillips, L. D.: 1984, A Theory of Requisite Decision Models, *Acta Psychologica*, **56**, 29–48.

Pirsig, R.: 1974, *Zen and the Art of Motorcycle Maintenance*, Bantam, New York.

Polanyi, M.: 1958, *Personal Knowledge*, Routledge and Kegan Paul, London.

Popper, K. R.: 1935, *The Logic of Scientific Discovery*, Huchinson, London (1959).

Prescott-Clarke, P.: 1982, *Public Attitudes towards Industrial, Work-Related and Other Risks*, Social Community Planning Research, London.

Purdy, M.: 1988, Guideline on Radon Misleading, *Philadelphia Inquirer*, 24 October.

Ravetz, J. R.: 1971, *Scientific Knowledge and Its Social Problems*, Oxford U.P.

Ravetz, J. R.: 1989, *The Merger of Knowledge with Power*, Mansell, London.

Ravetz, J. R., Macgill, S. M. and S. O. Funtowicz: 1986, Disasters Bring the Technological Wizards to Heel, *The Guardian*, 19 May, London.

Rivard, J. V. *et al.*: 1984, *Identification of Severe Accident Uncertainties*, NUREG/CR-3440, SAND 83–1689, U.S. Nuclear Regulatory Commission, Government Printing Office, Washington D.C.

Robinson, A.: 1966, *Non-Standard Analysis*, North Holland, Amsterdam.

Ruckelshaus, W. D.: 1984, Risk in a Free Society, *Risk Analysis*, **4**(3), 157–162.

Saaty, T. L.: 1980, *The Analytic Hierarchy Process*, McGraw-Hill, New York.

Savage, L. J.: 1954, *The Foundations of Statistics*, J. Wiley, New York.

Shapere, D.: 1986, External and Internal factors in the Development of Science, *Science and Technology Studies*, **4**(1), 1–9.

Sietmann, R.: 1987a, School Ranking Inconclusive, *The Scientist*, 15 June.

Sietmann, R.: 1987b. West Germans Debate Research Indicators, *The Scientist*, 29 June.

Spangler, M. B.: 1981, Risks and Psychic Costs of Alternative Energy Sources for Generating Electricity, *The Energy Journal*, **2**(1), 37–59.

Spencer, G.: 1987, Private Communication.

Stahl, W.: 1962, *Roman Science: Origins, Development and Influence to the Later Middle Ages*, University of Wisconsin Press.

Stephanou, H. E. and A. P. Sage: 1987, Perspectives on Imperfect Information Processing, *IEEE Transactions on Systems, Man and Cybernetics*, **SMC-17**(5), 780.

Stevens S. S.: 1946, On the Theory of Scales of Measurement, *Science*, **103**, 677–680.

Stutz, B.: 1988, Science for Sale, Ecologists call Colleagues 'Biostitutes', *The Scientist*, 28 November.

Taylor B. N., Parker, W. H. and D. N. Langenberg: 1969, *The Fundamental Constants and Quantum Electrodynamics*, Academic, London.

The Guardian: 1985, The Index Goes Haywire, London, 10 July.

The Royal Society: 1983, *Risk Assessment*, London.

Tiles, M.: 1984, *Bachelard: Science and Objectivity*, Cambridge U.P.

Thompson, M. and M. Warburton: 1985, Decision Making under Contradictory Certainties, *Journal of Applied Systems Analysis*, **12**, 3–34.

Thomson, W. A. R.: *Black's Medical Dictionary*, A. and C. Black, London.

Toulmin, S.: 1953, *The Philosophy of Science*, Hutchinson, London.

Tufte, E. R.: 1983, *The Visual Display of Quantitative Information*, Graphics Press, Cheshire (Conn.).

Turner, B.: 1978, *Man-Made Disasters*, Wykeham Press, London.

U.S. National Research Council: 1983, *Risk Assessment in the Federal Government; Managing the Process*, press release March 1, Washington D.C.

U.S. Nuclear Regulatory Commission: 1975, *The Reactor Safety Study – An Assessment of Accident Risks on U.S. Commercial Nuclear Power Plants*, NUREG-75/014, WASH-1400, Government Printing Office, Washington D.C.

van Heijenoort, J.: 1977, *From Frege to Gödel: A Sourcebook in Mathematical Logic*, Harvard U.P., Cambridge (MA).

Vesely, W. E. and D. M. Rasmuson: 1984, Uncertainties in Nuclear Probabilistic Risk Analyses, *Risk Analysis*, **4**, 313–322.

Warwick, A.: 1989, Private Communication.

Weinberg, A. M.: 1972, Science and Trans-Science, *Minerva*, **10**, 209–222.

Whittemore, A. S.: 1983, Facts and Values in Risk Assessment for Environmental Toxicants, *Risk Analysis*, **3**, 23–34.

Wilder, R. L.: 1968, *Evolution of Mathematical Concepts*, J. Wiley & Sons, New York.

World Health Organization: 1984a, *Behind the Statistics*, Geneva.

World Health Organization: 1984b, *The international Drinking Water Supply and Sanitation Decade: Review of National Baseline Data*, Geneva.

Wright, G.: 1984, *Behavioural Decision Theory*, Penguin, Harmondsworth, Chapter 3.

Zadeh, L. A.: 1965, Fuzzy Sets, *Information and Control*, **8**, 338.

Ziman, J. M.: 1960, Scientists: Gentlemen or Players, *The Listener*, **68**, 599–607.

INDEX

THEORY AND DECISION LIBRARY

SERIES A: PHILOSOPHY AND METHODOLOGY OF THE SOCIAL
SCIENCES

Already published:

Conscience: An Interdisciplinary View
Edited by Gerhard Zecha and Paul Weingartner
ISBN 90–277–2452–0

Cognitive Strategies in Stochastic Thinking
by Roland W. Scholz
ISBN 90–277–2454–7

Comparing Voting Systems
by Hannu Nurmi
ISBN 90–277–2600–0

Evolutionary Theory in Social Science
Edited by Michael Schmid and Franz M. Wuketits
ISBN 90–277–2612–4

The Metaphysics of Liberty
by Frank Forman
ISBN 0–7923–0080–7

Principia Economica
by Georges Bernard
ISBN 0–7923–0186–2

Towards a Strategic Management and Decision Technology
by John W. Sutherland
ISBN 0–7923–0245–1

Social Decision Methodology for Technological Projects
Edited by Charles Vlek and George Cvetkovich
ISBN 0–7923–0371–7

Reductionism and Systems Theory in the Life Sciences
Edited by Paul Hoyningen-Huene and Franz M. Wuketits
ISBN 0–7923–0375–X

Understanding Economic Behaviour
Edited by Klaus G. Grunert and Folke Ölander
ISBN 0–7923–0482–9

The Lifetime of a Durable Good
by Gerrit Antonides
ISBN 0–7923–0574–4